D0296262

An Appeal to Reason

Also by Nigel Lawson

The View from No. 11: Memoirs of a Tory Radical

The Power Game

The Nigel Lawson Diet Book

An Appeal to Reason

A Cool Look at Global Warming

Nigel Lawson

There is no opinion, however absurd, which men will not readily embrace as soon as they can be brought to the conviction that it is generally adopted.

(Schopenhauer, *Die Kunst Recht zu Behalten*)

Duckworth Overlook

London • New York • Woodstock

First published in 2008 by
Duckworth Overlook

LONDON
90-93 Cowcross Street, London EC1M 6BF
Tel: 020 7490 7300
Fax: 020 7490 0080
info@duckworth-publishers.co.uk
www.ducknet.co.uk

NEW YORK
141 Wooster Street, New York, NY 10012

WOODSTOCK
One Overlook Drive, Woodstock, NY 12498
www.overlookpress.com
[for individual orders and bulk sales in the United States,
please contact our Woodstock office]

A catalogue record for this book is available
from the British Library

ISBN 978-0-7156-3786-9 (UK)
ISBN 978-1-5902-0084-1 (US)

Typeset by Ray Davies
Printed and bound in Great Britain

To David Henderson,

who first aroused my interest in all this

Contents

Foreword & Acknowledgements

This is the fourth book of mine to appear. But whereas all four are very different in both genre and subject matter, this one is different in an additional way. While my three previous books had no difficulty whatever in finding a British publisher (indeed they did so before they were even written), this book, despite being promoted by an outstanding literary agent, was rejected by every British publisher to whom it was submitted – and there were a considerable number of them.

As one rejection letter put it: 'My fear, with this cogently argued book, is that it flies so much in the face of the prevailing orthodoxy that it would be very difficult to find a wide market'. The prevailing orthodoxy can be both stifling and intolerant. Those who have the temerity to question it have to become accustomed to being labelled 'deniers' – a loaded term – and of being accused of being in the pay of either 'big oil' or coal-mining interests. In the circumstances, I suppose I need to make it clear that I have not in fact received a penny from any commercial interest; although what I find most insulting is the implication that, if I had, that would determine what I write and say.

So, first and foremost, I would like to thank the distinguished American publisher, Peter Mayer, the proprietor of The Overlook Press in New York, who also owns the old-established London publishing house of Duckworth, for his courage in publishing this book. I must also thank my agent, Ed Victor, who had little idea of what a difficult task his would be when he agreed to take this on.

The origin of the book lies in a lecture I gave in London in

November 2006 under the auspices of the Centre for Policy Studies. It was clear at the time that something rather more substantial than a lecture was needed and I undertook to provide them with a pamphlet to publish. Once I had embarked on writing it, it soon became clear that something more substantial than a pamphlet was required: hence this book. I am grateful to the CPS, and in particular to its then Director, Ruth Lea, both for inviting me to give the lecture and for releasing me from my undertaking to provide them with a pamplet.

I must also thank Caroline Boin, of International Policy Network, who most efficiently and expeditiously compiled the bibliography and helped me with a number of the references and sources that appear in the notes at the back of the book. The notes also contain points of substance which might have interrupted the flow had they been in the main text, but which I hope will be read alongside it.

Finally, I must warmly thank a number of friends and family who took the trouble to interrupt their often very busy lives to read the book in draft and to let me have their comments. In strict alphabetical order, I am grateful to Samuel Brittan, Ian Byatt, Bob Carter, John Freeman, David Henderson, David Holland, Tony Jay, Roger Kerr, Mervyn King, Dominic Lawson, Tom Lawson, Dick Lindzen, Julian Morris, Andrew Tyrie and Tyrrell Young. Needless to say, none of them is in any way responsible for any of the views expressed in this short book, which are entirely my own.

Mourède, January 2008

Introduction

Over the past half-century we have become used to planetary scares of one kind or another. In the late 1960s, for example, the Malthusian nightmare re-emerged as we were authoritatively told that an unstoppable population explosion was inexorably leading to mass global starvation in the very near future.[1] A little later, we were warned by the Club of Rome, supported by large numbers of scientists and others, that the world was fast running out of natural resources and that, within our lifetimes, world economic growth would grind to a halt.[2] Then, during the early and mid-1970s, when the planet's temperature, which had been gently rising for most of the period since the so-called 'Little Ice Age' some 400 years ago, appeared to be falling again, many eminent scientists warned us that we were facing the disaster of a new ice age.[3]

But the latest scare – global warming – has engaged the political and opinion-forming classes to a greater extent than anything since Malthus warned us, a little over 200 years ago, that unless radical measures were taken to limit population growth, the world would run up against the limits of subsistence, leading inevitably to war, pestilence and famine.[4] This is perhaps partly because, at least in the richer countries of the world, we have rightly become more concerned with environmental issues. But that is no excuse for abandoning reason.[5] It is time to take a cool look at global warming.

By way of preamble, I readily admit that I am not a scientist. But then neither are the vast majority of those who pronounce on the matter with far greater certainty than I shall do here. Moreover (and this is frequently overlooked) the great majority of those scientists who speak with such certainty and apparent authority about global warming and climate change, are not in fact climate scientists, or indeed earth scientists, of any kind, and thus have no special knowledge to contribute.

Nor are those who have to take the key decisions about these critical issues scientists, let alone climatologists. Rather they are responsible politicians who, having listened to the opinions of scientists, must reach the best decisions they can in the light of the expert evidence available to them – just as I did, for example, in a not wholly unrelated field, when I was Energy Secretary in Margaret Thatcher's first government in the early 1980s.

More important still, science is only part of the story. Even if the climate scientists can tell us what is happening and why, they cannot tell us what governments should be doing about it. For this we also need an understanding of economics, in the sense of both economic forecasting (the likely growth of the world economy over the rest of this century, and how energy-intensive that growth is likely to be) and, even more importantly, economic analysis; what is the most cost-effective way of tackling the issue? And we also need an understanding of the politics; of what measures are politically realistic, a particularly tricky matter given the inescapably global nature of the issue. Finally, there is the ethical aspect, which is not as straightforward as it is usually made out to be.

In the subsequent chapters of this short book, I shall examine each of these dimensions of the global warming issue. I deliberately use the term 'global warming', rather than the attractively alliterative weasel words, 'climate change', throughout. This is because the climate changes all the time, it always has done and always will do, for reasons that may have little or nothing to do with temperature, let alone with man, and are only imperfectly understood. What is at issue is something much more specific: is the world getting warmer, if so why, how much warmer is it likely to get, what are the likely consequences and how much do they matter, and what can and should we do about it. To confuse global warming with climate change can lead the unwary to suppose that any significant or unusual weather event must be a consequence of global warming, which may very well not be the case.

First I shall examine the science, the extent to which it can be said to be settled, and what the climate scientists know or believe to be the case, as well as taking a look at the historical temperature record. In the following chapter I shall examine the prospect for the next hundred years if the conventional scientific wisdom is correct, and assess how serious is the threat to the planet. Chapter 3 explains the importance of taking fully into account mankind's ability to adapt to higher temperatures, in terms both of assessing the likely impact of any global warming that may occur, and of deciding the most cost-effective policy response.

Chapter 4, 'Apocalypse and Armageddon', looks at whether there are any specific disasters in the offing that should qualify the judgment reached at the end of Chapter 2. Chapter 5 looks at the possibility of reaching a global agreement on measures to mitigate global warming, while Chapter 6 looks at what the cost of such measures might be, and what form these measures might take. Chapter 7 compares the cost of taking action now with the benefits such action may confer, and looks at our approach to risk, uncertainty and to the important ethical dimension. The final chapter reaches conclusions about what we should rationally be doing about all this, and about why the issue of global warming has acquired the extraordinary salience it has.

I do not for a moment believe that this book will shake the faith of the true believers – that would be far too uncomfortable for them. But I suspect that most people have not yet made up their minds on this important issue, and may thus be amenable to reason. It is for them that this book has been written.

Chapter 1

The Science – and the History

It is frequently claimed by those who wish to stifle discussion, that the science of global warming is 'settled'. Even if it were, for the reasons I have already indicated – economic, political, and also ethical reasons – that would not be the end of the matter. But in fact, the science of global warming is far from settled.

This is, understandably, most unwelcome to our political leaders. It was reputedly Winston Churchill who demanded to be sent a one-handed economist; he was fed up with economists who, in answer to his questions, invariably replied: 'On the one hand ... but on the other hand ...' In the same way, it is one-handed scientists who most commend themselves to governments.

But the truth of the matter is that while some of the science is settled, there is much that is not. That is not to say that, even on the unsettled science, there is not a majority view – it could scarcely be otherwise – which can loosely be called the conventional wisdom. But the scope for uncertainty in this relatively new and highly complex branch of science is considerable – as indeed the House of Lords Select Committee on Economic Affairs discovered when it looked into the matter a few years ago, and explained in its unanimous all-party report.[1]

In any event, scientific truth is not established by counting heads.[2] There are many instances in the history of science in which subsequent evidence has overturned what had hitherto been the conventional wisdom. Nor, incidentally, does the fact that a scientific hypothesis has been published in a 'peer reviewed' learned journal provide *ipso facto* any evidence either that the science is 'settled', or that the hypothesis in question is likely to be proved correct. It does not even mean that the author's data and methods are available for scrutiny, or that his results are reproducible, as scientific journals, in contrast to most leading economic journals, do not require this. While peer review may be a useful process, all it means is that the author's peers consider that the paper which advances the hypothesis is worthy of publication in the journal to which it has been submitted. And it undoubtedly produces a bias in favour of whatever happens to be the conventional wisdom of the time.[3]

There is, indeed, a real question about the extent to which modern global warming science is genuine science at all. In the first place, as James Lovelock has pointed out: 'Observations and evidence are out of fashion; most evidence now is taken from the virtual world of computer models.' [4] In the second place, as Karl Popper long ago explained, for a theory or hypothesis to be genuinely scientific (rather than essentially metaphysical), its predictions have to be falsifiable by evidence in the real world.[5] It is not immediately apparent what real-world evidence could shake the faith of the true believers and overturn the conventional global warming wisdom, rather than at most causing its guardians to tweak their computer models.

Be that as it may, there is of course little doubt that, globally, the 20th century ended slightly warmer than it began. According to the world-renowned Hadley Centre for Climate Prediction and Research, an offshoot of Britain's Met Office:

> Although there is considerable year-to-year variability in annual-mean global temperature, an upward trend can be clearly seen [in the chart to which this commentary is appended]; firstly over the period from about 1920–1940, with little change or a small cooling from 1940–1975,

followed by a sustained rise over the three decades since then.[6]

This was published in 2005, but even by then the final claim was a trifle disingenuous, since what the chart actually showed (as has been confirmed by subsequent readings) was that the 'sustained rise' took place entirely during the last quarter of the last century. There has, in fact, been no further global warming since the turn of the century, although of course we are still seeing the consequences of the 20th century warming. The most recent global temperature series for the 21st century to date, published by the Hadley Centre in conjunction with the Climatic Research Unit of the University of East Anglia, runs as follows:

2001	0.40
2002	0.46
2003	0.46
2004	0.43
2005	0.48
2006	0.42
2007	0.41

The numbers represent degrees centigrade above the 1961–90 estimated global average temperature.

The figure for 1998, incidentally, was 0.52°C/0.9°F. The 21st century standstill (to date), which has occurred at a time when global CO_2 emissions have been rising faster than ever, is something that the conventional wisdom, and the computer models on which it relies, completely failed to predict.[7]

Indeed, since the statement by the Hadley Centre quoted above was written, the scientists there have significantly revised their views on this point.[8] Concluding that the climate models used hitherto have taken inadequate account of natural temperature variability, they have modified their model and now forecast that, after an unpredicted, almost decade-long lull, global warming will resume in 2009 or thereabouts. Maybe it will; we shall see. There are other scientists who suspect that the 21st century temperature standstill may be due to a marked observed decline in solar activity (of which sunspots are the most visible manifestation) after an exceptionally high level of solar activity in the last quarter of the 20th century.[9] If this is so, and if the current low level of solar activity continues, then, clearly, all bets are off. To repeat; we shall see.

The lull so far appears to be the average of a continued, if slower, gentle warming in the northern hemisphere (where most of us live) and an unpredicted slight cooling in the southern hemisphere. The Hadley Centre scientists now – after the event – seem to attribute this last to cooler water in the tropical Pacific Ocean, and an apparent resistance to warming in the Southern Ocean.

As a result of the absence of any recorded 21st century global warming, the formulation now favoured by the alarmists is that eleven of the last twelve years have been the warmest since records began. It is rather as if the world's population had stopped rising and all the demographers could say was that in eleven of the last twelve years the world population had been the highest ever recorded. (It is, incidentally, not hard to imagine what the serried ranks of climate alarmists would have been claiming had the global graph continued upwards, rather than stabilizing, during the early years of the present century.)

But even this may not be true, although we shall probably never know. Calculating the average global temperature is not as straightforward as it might appear at first sight. There are two problems: the first is how best to calculate the global average from the mass of data from individual weather stations around the world; the second is the reliability of the data, in particular that from much of the developing world and the former Soviet Union, notably during the period before there was such intense interest in the issue. It is worth, as a cross-check, taking a look at the data from the United States, where it is widely agreed, at least so far as economic statistics are concerned, that there is a greater degree of reliability than there is in many other parts of the world. And with its large land mass, the United States' temperature record ought to be a reasonable proxy for the Northern hemisphere as a whole.

The official US temperature history produced by NASA (and recently revised to correct computer errors) clearly shows both the mid-20th century cooling and the warming that occurred during the last quarter of the 20th century. But the cooling appears much more pronounced than that shown in the Hadley Centre's northern hemisphere chart. For the United States, only three of the last twelve years emerge as among the warmest since records began; and the warmest year of all was 1934.[10]

A further potential complication when using the data from the temperature records is caused by urbanization. It is settled science that urbanization raises near-surface temperatures; this is known to aficionados as the 'urban heat island' effect. It was first observed in London in the early 19th century[11], and can now be measured using satellite infrared imagery. There are two problems which arise from this. The first is the obvious question of how much of the recorded global warming has in fact been caused by this process which, although man-made, is wholly unrelated to greenhouse gas emissions. The second and probably greater problem, is the extent to which the recorded rise in surface temperatures in the late 20th century has been exaggerated by the fact that a large proportion of climate stations are located in cities or at airports on the developing edges of cities – areas which have become increasingly urbanized over the past hundred years.[12]

Apart from the trends, there is of course the matter of the absolute numbers. The Hadley Centre chart suggests that, for the first phase, from 1920 to 1940, the increase was 0.4ºC/0.7ºF. From 1940 to 1975 there was a cooling of about 0.2ºC/0.4ºF.[13] Finally, since 1975 there has been a further warming of about

0.5ºC/0.9ºF, making a total increase of some 0.7ºC/1.3ºF over the 20th century as a whole (from 1900 to 1920 there was no net change).[14] Thus a modest – if somewhat intermittent – degree of global warming has evidently occurred. Why has this happened?

It is popularly supposed by politicians and the media that scientists have demonstrated that the sole cause of global warming is the growth in man-made carbon dioxide (CO_2) emissions, which began with the industrial revolution in the 19th century and has continued ever since.[15] But in fact the science is both much more complex than this, and much less certain. So let us begin with what is, to all intents and purposes, certain.

First, there is no doubt that atmospheric concentrations of carbon dioxide increased substantially during the 20th century – by more than 30% in fact – and indeed have continued to do so throughout the 21st century lull in global warming. There is little dispute that this increase is very largely man-made – a result of the carbon emissions caused by the rapid worldwide growth of carbon-based energy consumption (the burning first of coal, and subsequently of oil and gas).

It is also settled science that carbon dioxide (the great bulk of which, incidentally, is not man-made but natural) is one of a number of so-called 'greenhouse gases', whose combined effect in the earth's atmosphere is to keep the planet significantly warmer than it would otherwise be. To oversimplify, most of the heat the earth receives from the sun bounces back into space in the form of infrared radiation. But the so-called greenhouse gases in the earth's atmosphere prevent some of this infrared radiation, and thus some of the heat, from escaping back into space. Without this, life as we know it would be unable to survive. Far and away the most important of these gases – thought to account for at least two-thirds of the greenhouse effect – is water vapour, including water suspended in clouds. Rather a long way behind, carbon dioxide is the second most important greenhouse gas.[16]

It is worth noting at this point that to describe the carbon dioxide in the atmosphere as pollution is as absurd as it would be to describe the clouds as pollution. Those who do so are misrepresenting the science in order to mislead. There are, of course, noxious substances in the atmosphere (such as sulphate aerosols, about which more later), and the more we can do to eliminate them, the better. But carbon dioxide is not one of them. It is in fact a life force; just as we (and all other animals) need to breathe oxygen in order to survive, so plants (of all kinds) need to breathe carbon dioxide to survive, and in general, the more carbon dioxide there is in the atmosphere, the better the development of plant life on the planet. This is customarily referred to as the 'fertilization effect' of CO_2, and is scarcely a bad thing.

Given the greenhouse effect, it can also be said to be settled science that the marked, and largely man-made, increase in carbon dioxide concentrations in the atmosphere has contributed to the modest 20th century warming of the planet. But what is far more complex, and far less certain, is *how great* a contribution it has made. Clearly the answer to this question has a direct bearing both on predictions of future global temperatures, and on the efficacy and cost of any measures we may wish to take to try to prevent excessive warming. And it is of course these predictions, and not the modest amount of warming that has so far occurred, that give rise to concern.

It was to advise governments on these and related issues, such as the likely practical impact of any future temperature rise, that the Intergovernmental Panel on Climate Change (IPCC) was set up in 1988, under the auspices of the United Nations. Essentially the IPCC neither carries out research, nor monitors the climate; its purpose is to assimilate everything that is being done in this field, and to provide a synthesis to inform the governments of the world.

Despite the very real dangers inherent in giving any organization a global quasi-monopoly of official scientific (or other) advice, on balance this was in principle a sensible step. Unfortunately, in a number of important respects, the IPCC's processes have become seriously flawed. This was one of the major findings of the House of Lords Report to which I have already referred, and the flaws have been more fully exposed in a number of subsequent studies.[17]

The problem stems in part from the fact that what was intended by the governments who set up and continue to fund the IPCC to be a fact-finding and analytical exercise, has mutated in the minds of most of those who head it into something more like a politically correct alarmist pressure group.

For example, the *Financial Times*' report of the press conference in Paris on 2 February 2007, which launched the publication of the first instalment of the IPCC's most recent report, opened with the words: 'Only urgent international action to cut emissions can prevent climate-related catastrophe, scientists warned on Friday.'[18] In fact, the report (and its scientists) says nothing of the sort. But this was almost certainly an accurate summary of the statements made by Mr Achim Steiner, director general of the UN Environmental Programme, Mr Yvo de Boer, secretary general of the UN Framework Convention on Climate Change, and Mr Rajendra Pachauri, the IPCC's chairman, who together gave the press conference.

Nevertheless, despite these damaging flaws, the IPCC does some useful work, and its reports constitute the conventional wisdom (or majority view) on all the key global warming issues. There are, of course, innumerable other voices proclaiming this conventional wisdom, but the IPCC is far and away the most authoritative and influential, and any serious discussion of the subject has to address, as I shall therefore do in this book, the account the IPCC has presented to the world's governments and media.

On the central question of how much of the modest observed 20th century global warming can be attributed to man-made (or in IPPC parlance, for the avoidance of apparent gender discrimination, 'anthropogenic') emissions, its most recent report[19] argues that:

> The observed widespread warming of the atmosphere and ocean ... support the conclusion that it is ... very likely that it is not due to known natural causes alone...

and concludes that:

> *Most* [the emphasis is mine] of the observed increase in globally averaged temperatures since the mid 20th century is very likely due to the observed increase in anthropogenic greenhouse gas concentrations.[20]

While the first of these statements is more or less unexceptionable, the second is strongly disputed by many reputable climate scientists.[21] It may of course be correct, and we need to bear in mind that 'most' simply means more than 50% of the 0.5ºC/ 0.9ºF warming since the 1950s; but then again, it may not be. Some of the sources of uncertainty are worth spelling out.

The first is that the science of clouds, which is clearly critical (not least because water vapour, as we have seen, is far and away the most important contributor to the greenhouse effect), is one of the least understood aspects of climate science. This includes the important question of the interaction between clouds and other greenhouse gases, in particular carbon dioxide, and also between clouds and higher temperatures. Indeed, it is these conjectural feedbacks which largely determine the computer models on which, as we will see later, conventional climate prediction rests. Most existing climate models employed to predict future temperature levels treat clouds in a way that amplifies the warming effect of carbon dioxide, but this treatment is disputed.[22]

As the IPCC has acknowledged: 'Cloud feedbacks remain

the largest source of uncertainty.'[23] Or, in the candid words of the Hadley Centre: 'There are many of these feedbacks, both positive and negative, many of which we do not fully understand. This lack of understanding is the main cause of the uncertainty in climate predictions; this applies in particular to changes in clouds.'[24] The uncertainties over cloud feedbacks have in fact been recognized for the past 30 years or so, during which time, despite the best efforts of the climate scientists, there has been remarkably little progress in resolving them.

Another uncertainty in attributing the cause of the slight 20th century warming derives from the fact that, while the growth in man-made carbon dioxide emissions, and thus carbon dioxide concentrations in the atmosphere, continued relentlessly during the last century, and continues unabated to this day, the global mean surface temperature, as we have already seen, has increased in fits and starts. In particular, there was, in the words of the Hadley Centre: 'little change or a small cooling from 1940–1975'.[25]

The conventional wisdom, duly incorporated in the immensely complex computer models developed by the Hadley Centre and others, which form the basis of the IPCC's conclusions and which generate a specific temperature rise for any projected increase in carbon dioxide emissions and concentrations, is that this must have been due to the fact that, before 1975, power stations (largely coal-fired) emitted large amounts of sulphate aerosols into the atmosphere, which – by dimming the rays of the sun – more than offset the warming effect of the increased carbon dioxide. Since 1975, it is claimed, when the industrialized West took steps to reduce this pollution, the carbon dioxide effect has reigned supreme.

This may of course be so. But it is pure conjecture. While aerosols in the atmosphere do indeed cool the planet to some extent, leading scientists in this field readily concede that the extent to which they do so is unclear. Indeed, they claim that as good a way as any to establish this is to calculate the amount of aerosol 'forcing' needed to cancel out the carbon dioxide 'forcing' over this period – which of course makes the whole argument completely circular.[26] The Hadley Centre explicitly concedes that it is possible that, in their model, 'the heating effect of man-made greenhouse gases and the cooling effect of man-made aerosols have [both] been overestimated'.[27]

Another problem with the aerosol cooling theory, which is an integral and quantified part of the Hadley Centre model, is that it makes it difficult to explain as man-made the marked warming that occurred between 1920 and 1940, when power station emissions were as dirty as they have ever been. The Hadley Centre claims that observed increased solar radiation may explain some 0.1ºC/0.2ºF of the 0.4ºC/0.7ºF rise in global mean temperature over that period, but it still leaves the greater part unexplained.[28] And if unexplained natural variation is the only possible cause of the bulk of the warming between 1920 and 1940, why may it not have been a major factor at other times?

A further cause of uncertainty is the problem of the tropical troposphere, the atmosphere between the tropics of Cancer and Capricorn up to 15 kilometres in altitude. As the Hadley Centre concedes,[29] 'analysis of satellite data shows substantially less warming than at the surface. Climate models predict that we should have seen a relatively greater warming in the troposphere than at the surface; this potential discrepancy between models and observations is not well understood, although uncertainty in observations is the more likely explanation.' Alternatively, the models might just be wrong, or else the data may be recording warming that is not greenhouse-related.

Since the Hadley Centre booklet appeared, further work has been done on this by the US Government's highly-regarded Climate Change Science Program (CCSP). Some climate scientists claim that this has reconciled the disparity between the observations and the models, while others believe that it has done nothing of the sort. The point is clearly an important one, as it bears directly on the question of whether the man-made greenhouse effect is the major cause of the late-20th century warming or only a minor one.[30]

Before leaving the vexed question of the Hadley Centre and other, similar, climate models on which the IPCC relies, it is worth noting the recent acknowledgement by Dr Kevin Trenberth that 'None of the models used by the IPCC are initialized to the observed state and none of the climate states in the models correspond even remotely to the current observed climate.'[31] The New Zealand-born Dr Trenberth, head of the Climate Analysis Section of the US National Center for Atmospheric Research, is no sceptic; he is a distinguished pillar of the IPCC (he was one of the principal authors of the scientific section of its 2007 Report) and a committed adherent of the conventional climate change wisdom.

But enough of models. What about the real world? Another great source of uncertainty in all this is that the earth's climate has always been subject to natural variation which has been wholly unrelated to man's activities. Climate scientists differ about the causes of this, although many believe that variations in solar activity have played a key part.[32] Variations in ocean currents can also have a marked effect on global temperatures.

Although before the middle of the 19th century there were few reliable scientific temperature measurements, it is well established, for example, from historical records and accounts, that a thousand years ago, well before the onset of industrialization, there was what has become known as the mediaeval warm period, a benign time when temperatures were probably at least as high as, if not higher than, they are today.[33] Going back even further, during the Roman Empire, it was probably even warmer. There is archaeological evidence that in Roman Britain, for example, vineyards existed on a commercial scale at least as far north as Northamptonshire.[34]

More recently, during the 17th and early 18th centuries, there was what has become known as the Little Ice Age, when the

Thames was regularly frozen over in winter and substantial ice fairs held on the frozen river became a popular attraction, immortalized in colourful contemporary prints. Treeline studies showing how far up mountains trees are able to grow at different historical periods – something which is likely to be correlated with temperature change – suggest that these temperature variations, in particular the mediaeval warm period, occurred outside Europe as well.[35] Recent ocean sediment studies appear to confirm this.[36]

A very different account of the past was given by the now notorious 'hockey-stick' chart of global temperatures over the past 1,000 years, so called because the allegedly constant temperature over the long period before reliable scientific observations were first recorded (so much for the well-authenticated mediaeval warm period) resembled the straight handle, and the subsequent 20th century rise the curved blade, of a hockey stick. Produced by Professor Michael Mann and his colleagues, Bradley and Hughes, in the United States in 1998[37], and prominently displayed by the IPCC in its 2001 Report, it rapidly became the icon of the conventional global warming wisdom.

Its purpose, of course, was to illustrate how it was clearly industrialization that must have caused the 20th century warming, and it was uncritically embraced by all those subscribing to this view – among them the British government, which also featured the hockey stick chart prominently in its 2003 energy White Paper. Indeed, had it not been for a penetrating critique painstakingly developed by two Canadians – Professor Ross McKitrick, an environmental economist specializing in global warming issues, and Steve McIntyre, a minerals consultant with a particular expertise in statistics, the hockey stick myth would have become the received wisdom, as the IPCC evidently wished it to be.

McIntyre and McKitrick had been irked not least by Mann's refusal to share his data or divulge his methods. This is, regrettably, an all too familiar characteristic of the global warming community. Indeed, it applies even to the published mean global temperature series itself, which is produced by the Climate Research Unit at the University of East Anglia. When an Australian researcher asked the CRU to let him examine the underlying data and methods used to compile the series, Dr Jones of the CRU refused, writing: 'Why should I make the data available to you, when your aim is to try and find something wrong with it?'[38]

However, in the light of the Canadian critique, two US congressional committees decided to set up two committees of experts, one of climate scientists and the other of statisticians, to review the matter. They reported in 2006, and, taken together, their reports wholly vindicated McKitrick and McIntyre's critique. The 'hockey stick', which the IPCC had uncritically endorsed, was shown to be an artefact of inadequate and selective sampling and improper statistical methodology, and is now comprehensively discredited.[39] All that remains is an acknowledgement that 'it can be said with a high level of confidence that global mean surface temperature was higher during the last few decades of the 20th century than during any comparable period during the preceding four centuries'[40] – hardly a remarkable conclusion, given that 400 years ago saw the onset of the rigours of the Little Ice Age.

Regrettably, the IPCC has refused either to acknowledge the 'hockey stick' error or to review, let alone change, the slipshod procedures which allowed it to endorse it in the first place. All it has done – and in the circumstances it could scarcely do less – is quietly drop the hockey stick from its latest, 2007 Report.[41] Among the problems that have come to light is the fact that the methods used by Professor Mann generate hockey stick shapes from purely random data (what the scientists term 'red noise'). If (an elementary step, one would have thought) the proxy method he used to estimate the historical temperature record is compared with the actual 20th century record, for which we have direct measurements, it turns out that there is little similarity either for the period before 1900, or for that after 1950 (when, of course, the warming now at issue occurred).

But it is not just over time that the earth's climate displays considerable natural variability. Not only do both the temperature and the climate in general vary greatly across the globe, but both the amount of warming and its alleged consequences appear to vary geographically, too. For example, the recorded warming has been greatest in the polar regions, where it is coldest, and least in the tropics, where it is hottest – and, indeed, in the regions that have warmed the most, there is a pronounced seasonal difference, with the coldest months showing the greatest warming.[42] Again, there are parts of the world where glaciers are retreating, and others where they are not – and indeed some where they may even be advancing.[43] The fringes of the Greenland ice sheet appear to be melting, while at its centre, the ice is thickening.[44] Curiously enough, there are places where sea levels are perceptibly rising, and others where they are static, or even falling (partly, but by no means entirely, because the land itself is rising), suggesting that local factors still dominate any global warming effect on sea levels.[45]

This diversity makes it all too easy for the likes of Al Gore, as in his tendentious film *An Inconvenient Truth*, to cherry-pick local phenomena which best illustrate their predetermined alarmist global narrative. We need to stick firmly to the central point: what has been the rise in global mean temperatures over the past hundred years; why we believe this has occurred; how much, on this basis, are temperatures likely to rise over the next hundred years; and what are the consequences likely to be. It is only after answering these questions that we can begin to decide rationally what can or should be done about global warming.

There are two other points to be borne in mind:

The first is that, in the current state of global warming alarmism, there is a tendency, notably in the media, to attribute any and every uncomfortable weather event to global warming, and to regard this as proof of the need to curb carbon dioxide emissions. In fact, many of these events are by no means unprecedented, although of course the consequences may be; as populations rise and wealth increases, the cost of the damage caused by an adverse weather event is likely to be considerably greater, even in real terms, than that caused by a similar event many decades earlier. Again, many such events are purely local in origin, unrelated to any global trend, let alone global warming (and, of course, even those events which are thought to be at least in part a *consequence* of global warming, clearly tell us nothing about the *causes* of global warming).

The second is that our understanding of the climate is still, relatively speaking, in its infancy. Not only is the climate an immensely complex system, but climatology is a relatively new branch of science, having emerged from a variety of disciplines, notably meteorology, but also including oceanography and geology, during the course of the 20th century. It is understandable that many scientists, especially those engaged in advising governments, should tend to emphasize what they know or believe they know, notably the nature of the greenhouse effect, at the expense of what they do not know. But neither scientists nor politicians serve either the truth or the people by pretending to know more than they do. It is clearly illegitimate to assume that what is not at present known cannot exist. Uncertainty is uncertainty.

Few serious scientists, however, would descend to the level of superficiality that characterizes the statement in the UK government's Stern Review that 'The accuracy of climate predictions is limited by computing power … It is important to continue the active research and development of more powerful climate models to reduce the remaining uncertainties in climate predictions.'[46] As even the IPCC and the Hadley Centre concede, the 'remaining uncertainties' include key aspects of the science.[47]

I have to say that in my time as Chancellor, I would have

asked the Treasury to conduct a thorough assessment of anything as important as the economics of climate change *before* the government made any costly policy commitments. The Stern Review however, originally published by the UK government in 2006 (Sir Nicholas Stern was at the time Head of the Government Economic Service), was only embarked upon after the event, and is essentially a propaganda exercise in support of the UK government's predetermined policy of seeking a world leadership role on climate change. As a result, while its 692 pages contain much of interest, neither its conclusions nor the arguments on which they are based possess much merit.[48]

In one sense however, the difference over the science of global warming is relatively narrow. The conventional wisdom, as we have seen, is that most of the warming that occurred during the last quarter of the 20th century was very likely due to the growth of carbon dioxide emissions, and thus of CO_2 concentrations in the atmosphere as a result of industrialization. The opposing view is that, while this was almost certainly a factor, the use of the word 'most' is questionable. But the practical consequences of this difference are considerable. In the first place, they profoundly affect any predictions we may seek to make about the next hundred years. And in the second place, and arising from that, while the cost of cutting back sharply on carbon dioxide emissions is the same in either case, the benefits are clearly not.

As is frequently the case in government – notably, for example, in the field of national defence – we are faced with the task of deciding, rationally, the most sensible policy course to take against a background of fundamental uncertainty. We need to avoid being paralysed into complete inactivity. But we also need to avoid being panicked into what could be disastrously damaging action. In many ways the decision to invade Iraq is an instructive parallel (with the government's alarmist Stern Review playing the role of Mr Blair's notorious 'dodgy dossier').[49]

It is worth noting that, in the field of economic policy, it is nowadays generally accepted by central bankers that endemic uncertainty needs to be recognized as the central fact of life, and that as a result, policy responses should be chosen that make sense in terms of a whole range of possible scenarios. The origins of climate uncertainty are of course different from those of economic uncertainty, but the lessons for rational policy-making are the same.

It is, however, prudent to err on the side of caution. For that reason, in the remainder of this book I shall work on the assumption that the majority (IPCC) view described earlier in this chapter is correct, while bearing in mind that, to a very important extent, this issue is in fact anything but 'settled'.

Chapter 2

The Next Hundred Years: How Warm? How Bad?

There is something inherently absurd about the conceit that we can have any useful idea of what the world will be like in a hundred years time, and it is even more ridiculous to believe that computer modelling can open a window to the future that is otherwise closed. We have only to ask ourselves whether the Edwardians, even if equipped with the most powerful modern computers, would have been able to foresee the massive economic, political and technological changes that have occurred over the past hundred years.

Yet that is what the IPCC purports to do, and is the basis on which it provides its advice to the governments of the world, to the admiring attention of the political and media classes. So there is nothing for it but to take a look at how it sets about its task, so that we may arrive at an informed view of the results.

In a nutshell, to get a line on how much global warming there is likely to be over the next hundred years, and what the practical impact of the consequent rise in global temperatures might be, the IPCC adds to the assumed nature of the link between atmospheric concentrations of carbon dioxide and temperature its estimates of how much CO_2 emissions are likely in fact to increase over the next hundred years, and then assesses, largely in quantified form, the likely consequences of the resulting rise in world temperature.

Not that the IPCC's 2007 Report confines itself to looking ahead a mere hundred years. It is replete with references to 'estimated multi-century warming' and the consequences this might bring, and indeed its most alarming possible outcomes 'are projected to occur over millennial time scales'. But to take one small part of the overall climate picture, and then, on the basis of conjectural computer models, to project that forward indefinitely while assuming that everything else remains unchanged, is palpably absurd.[1] It is always essential to avoid the mistake of confusing the unknown with the unimportant. In any event, the very idea that we can either usefully or intelligently look a millennium or more ahead, as a basis for serious policy decisions, is farcical.

Indeed, even looking ahead a hundred years, the IPCC's principal time horizon, is clearly a very uncertain business. As Sir John Houghton, a former Chief Executive of Britain's Met Office, former Chairman of the Scientific Working Group of the IPCC, and one of the foremost pillars of the conventional climate change wisdom, has conceded: 'when you put models together which are climate models added to impact models added to economic models, then you have to be very wary indeed of the answers you are getting, and how realistic they are.'[2] The IPCC itself has described it as a 'cascade of uncertainty'.[3]

In slightly more colloquial terms, it might be said that you start with the uncertainties of long-range weather forecasting, add to these the uncertainties of long-range economic forecasting, plus the uncertainties of long-range population forecasting, feed them all into a powerful computer and supposedly arrive at a sound basis for serious – and seriously expensive – long term policy decisions.

So let us take a look at the first step in the process, the IPCC's projections of how much carbon dioxide emissions – in the absence of special measures to curb them – may rise over the next hundred years.

Its 'Special Report on Emissions Scenarios' (SRES) was produced in 2000.[4] This comprised six different scenario groups, all of which it claims should be considered equally sound, but which produce different rates of growth of emissions over the next hundred years. The differences arise largely because of different assumptions about economic growth, different assumptions about the energy-intensiveness of that growth, different assumptions about technological advance, and different assumptions about population growth.

But although they are different, the IPCC's scenarios do have one important feature in common: they all assume that, over the present century, faster economic growth will mean that living standards in the developing world, in the conventional sense of GDP per head of population, will to a very considerable extent catch up with living standards in the developed world. Only in the most pessimistic of the six scenarios (which needless to say is the one – A2 – which the Stern Review takes as its base case) are living standards in the developing world projected to be lower by the end of this century than they are in the developed world today. And even in this scenario, they are projected to rise to 75% of living standards in the developed world today, compared to a much smaller fraction at the present time. In all the other five scenarios, living standards in the developing world by the end of this century are projected to be significantly higher than they are in the developed world today – in the most optimistic scenario, more than three times as high.

In other words, by 2100 poverty really has, at least to a considerable and gratifying extent, become history. If nothing else, this should cheer up those who have been told that disaster stares us in the face if we do not take urgent action to save the planet.

It is only fair to add that what I have just spelled out is what emerges from the IPCC's scenarios before deducting the projected costs to the economy of 21st century global warming. I will, of course, come to that; and it will be seen that it does not fundamentally change the picture. It is also of course true that the IPCC's projections of 21st century economic growth may prove to have been too optimistic, but in this case, given the assumed growth–emissions–temperature nexus, there would be less global warming, too.

The SRES, which was produced in 2000 for the IPCC's 2001 Report, was also fundamentally flawed by the fact that all its global estimates were derived by taking national figures and converting them using market exchange rates instead of purchasing power parity, as all reputable economists and international institutions (such as the International Monetary Fund) insist on doing.[5] It was hoped, therefore that an updated and revised SRES, which would take into account both the exchange rate error and changes in the real world during the intervening years, would be prepared in time for its latest, 2007 report. The IPCC refused to do this. Its apologists, such as Dr Trenberth, airily explain that:

> There are no predictions by the IPCC at all. And there never have been. The IPCC instead proffers 'what if' projections of future climate that correspond to certain emissions scenarios. There are a number of assumptions that go into these emissions scenarios. They are intended to cover a range of possible self-consistent 'story lines'.[6]

Well, well. In fact, it is hard to escape the suspicion that the IPCC declined to update its emissions scenarios for fear that the new ones might give rise to lower temperature projections.

As it is, the temperature projections it does come up with on this somewhat suspect basis in its fourth and latest report, range from a rise in the global average temperature by the year 2100 of 1.8ºC/3.2ºF ('best estimate') for its lowest emissions scenario, to one of 4ºC/7.2ºF ('best estimate') for its highest emissions scenario, with a mean increase of under 3ºC/5.4ºF.[7] These projections were made and published, incidentally, well before the Hadley Centre acknowledged the early 21st century global warming lull, and amended its model to take account of this.

At this point it might be a good idea to leave the rarefied world of the IPCC for a moment and take a brief reality check.

A warming of 3ºC/5.4ºF over the next hundred years amounts to a warming of 0.03ºC/0.05ºF a year. In the last quarter of the last century (1975–2000) we experienced a warming of 0.02ºC/0.04ºF a year – not an enormous difference in absolute terms. During those 25 years of gentle warming the world managed pretty well. As Indur Goklany has pointed out, the United Nations compiles what it calls a 'human development index' (HDI), which is a combination of life expectancy, level of education and level of economic development. Its annual Human Development Report shows that all but three of the 102 countries for which data are available enjoyed an improvement in HDI between 1975 and 2002 (the three exceptions were the Democratic Republic of Congo, Zambia and Zimbabwe.) Goklany also shows that there was a similar near-universal improvement in measures of hunger and infant mortality.[8]

Again, is it really plausible that there is an ideal average world temperature, which by some happy chance has recently been visited on us, from which small departures in either direction would spell disaster? Moreover, while a sudden change in temperature might indeed be disruptive, what is at issue here is the prospect of a very gradual change over a hundred years and more.

In any case, average world temperature is simply a statistical artefact. The actual experienced temperature varies not only between day and night and between summer and winter. It also varies enormously in different parts of the globe; and man, whose greatest quality is his adaptability, has successfully colonized most of it. Two countries at different ends of the earth, both of which are generally considered to be economic success stories, are Finland and Singapore. The average annual temperature in Helsinki is less than 5ºC/41ºF. That in Singapore is in excess of 27ºC/81ºF – a difference of more than 22ºC/40ºF. If man can successfully cope with that, it is not immediately apparent why he should not be able to adapt to a change of 3ºC/5.4ºF, when he is given a hundred years in which to do so.

In a sense we adapt to the outdoor temperature in large part by mitigating the indoor temperature. We shall, of course, need to look seriously at why the IPCC, and the climate alarmists of whom it has more or less become the institutional embodiment, believe that there is indeed an existential threat. But when doing so it will be useful to keep these common-sense considerations in mind.[9]

The IPCC seeks to assess the likely impact of projected global warming over the next hundred years in two ways.[10] First, it looks separately at five major headings: *water, ecosystems, food, coasts* and *health*.[11] Then it adds all these impacts together to provide an overall figure of the cost to the world of the projected warming.

This last is of course intended to be the net cost. It is clear that while warming brings costs, it also brings benefits. Given the wide geographical variation in temperatures around the world, it is obviously likely that, while in the warmer regions the costs could be expected to exceed the benefits, in the colder regions the benefits might well exceed the costs. Happily, as we saw in the previous chapter, during the late 20th century warming, it was the hotter regions of the world that warmed the least, and the colder regions the most. The IPCC Report claims to take into account both costs and benefits, yet it devotes large amounts of space to the costs and almost none to the benefits.[12] It is difficult not to sense a lack of even-handedness, leading to a bias in the overall assessment.

But let us first take a look at each of the IPCC's five impact headings, in turn. The first is *water*. Certainly, the modest warming that has occurred so far has caused no shortage of water; it is estimated that over the 20th century as a whole there has been something like a 3% increase in the amount of water flowing down the world's rivers.[13] The IPCC is, however, concerned that there may be greater variability of water supplies, both regionally and seasonally. If so, the remedy is plain. Where and when water is in abundance, flood defences need to be constructed to cope with the wet season, as do water storage facilities for use during the dry season. Where there is shortage, as there was in many places long before there was any hint of global warming, the remedy is to cut out the widespread wasteful use of water through sensible water conservation measures, including in particular the pricing of water.[14]

In fairness, it should be added that those who see an increasing shortage of fresh water in many parts of the world as one of the most serious problems of the next hundred years, leading not only to increased drought, but to large-scale enforced population movements and perhaps even 'water wars', may have a point, but that point has nothing whatever to do with global warming. The problem is the huge increase in the world's population – currently some 6.6 billion and rising fast, although the UN's population experts expect it to stabilize at around 9.1 billion towards the end of this century – leading to a massive increase in the demand for fresh water, without any corresponding increase in the effective supply.

This does mean that improved water resource management, and above all the proper pricing of water, are of the first importance. Moreover, there has recently been encouraging progress in desalination technology, offering the promise of producing ample supplies of fresh water from the salt water of the sea. The successful genetic modification of plants may also have an important part to play in this context. But what is abundantly clear is that cutting back on carbon dioxide emissions is irrelevant. If one or more of these direct solutions to the problem can be implemented it will be otiose, and if (as is fortunately unlikely) none of them can be, it will, at best, merely scratch the surface of what is undoubtedly a major challenge.

The second of the IPCC's impact categories is *ecosystems*, where it states that 'approximately 20–30% of plant and animal species assessed so far are likely to be at increased risk of extinction if increases in global average temperature exceed 1.5–2.5°C/2.7–4.5°F (*medium confidence*)'. 'Medium confidence' (the italics are the IPCC's) is officially defined by the IPCC as meaning that the statement has about 5 chances out of 10 of being correct: in other words, even the IPCC regards this prediction as just as likely to be wrong as to be right. Moreover, as is well known, those animal species at risk of extinction are threatened far more by other factors than they are by warming.[15] It is hard to believe that anyone would seriously seek to justify drastic curbs in CO_2 emissions on these flimsy grounds.

This is particularly so since it is generally accepted by serious ecologists (that is, the scientists rather than the campaigners) that, over the past two-and-a-half-million years, a period during which the planet's climate fluctuated substantially, remarkably few of the earth's millions of plant and animal species became extinct.[16] This applies not least, incidentally, to polar bears, which have been around for millennia, during which there is ample evidence that polar temperatures have varied considerably.[17] The principal threat to polar bears, the population of which has in fact risen significantly over the past forty years, is hunting.[18]

The third category is *food*. This is obviously of the first importance. What the IPCC has to say here has not been widely reported: 'Globally, the potential for food production is projected to *increase* [my emphasis] with increases in local average temperature over a range of 1–3°C/1.8–5.4°F, but above that it is projected to decrease.' It will be recalled that the mean temperature increase suggested by the IPCC's various scenarios for the end of the present century is less than 3°C/5.4°F.

Moreover this is an area where the scope for adaptation is particularly pronounced. It is not simply a matter of farmers being able to make better use of irrigation and fertilizers, and indeed to switch to strains or crops better suited to warmer climes, should the need arise – something, incidentally, which will happen autonomously, without any need for government intervention – it is also because we are in the early stages of a revolution in agricultural technology, with the development of bio-engineering and genetic modification.

It is true that there is an irrational hostility to these developments in much of Europe, even though they are essentially modern versions of the centuries-old hit-and-miss genetic modification involved in selective breeding. But Americans have been eating new-style genetically modified food for well over a decade, and if there had been any adverse consequences these would have come to light long ago. In any event, this upmarket hostility is unlikely to be an impediment in the poorer countries where it really matters. The old Malthusian nightmare of mass starvation unless population growth was severely restricted arose from a prediction of the capacity for food production based on the technology of the time. It was the subsequent development of agricultural technology that made a nonsense of it. The parallel is an illuminating one.

The fourth impact category is *coasts*, where the IPCC is concerned about sea-level rise, brought about by a combination of ocean warming expanding the volume of water, and some melting of the Greenland and West Antarctic ice sheets, causing coastal flooding in low-lying areas. Sea levels have, in fact, been rising very gradually for as long as records exist, and there is little sign of any acceleration so far. Indeed, the most recent study[19] suggests that the average annual rise may have been slightly less in the second half of the 20th century than in the first half (1.5 millimetres a year against 2 millimetres a year).

Despite this, particular concern has been widely expressed for the fate of those who live in low-lying island states such as the Maldives in the Indian Ocean, and the small Polynesian islands (such as Tuvalu) in the South Pacific. In fact, studies have shown that the sea level in the Maldives has, if anything, been falling over the past thirty years.[20] As for Tuvalu (whose inhabitants, according to Mr Gore, are already fleeing to New Zealand because of sea-level rise: in fact they are economic migrants), there is no record of sea level before 1978. Since then, the tide-gauge that was put in place has recorded an annual sea-level rise of a negligible 0.7 millimetres a year. In 1993, scientists from Flinders University in Australia, believing that the old (float-type) tide-gauges must be inaccurate, set up new modern ones in a dozen Pacific islands, including Tuvalu. After more than a decade of finding no sign of any significant sea-level rise (in 2006 Tuvalu actually recorded a fall) the project has recently been abandoned.[21]

For the 21st century the IPCC projects a possible acceleration (yet to be recorded) in the trivial overall rate of sea-level rise that occurred during the 20th century, to anything between 1.8 millimetres and 5.9 millimetres a year, depending on the amount of warming – that is to say, a total of between 18 and 59 centimetres by the year 2100 (a far cry, incidentally, from the 20-foot rise that was one of the more startling images in Mr Gore's fanciful film). [22] Given our capacity to adapt to gradual change, a sea-level rise of, at most, less than a-quarter-of-an-inch a year is not, frankly, on a scale to be alarmed about.

The fifth, and last of the IPCC's impact categories is *health*. In its 2001 Report, the IPCC focused heavily on a projected increase in the incidence of malaria brought on by warming. However malaria experts pointed out that, in fact, temperature has little bearing on the spread of the disease, which is mainly caused by other factors altogether. Indeed, malaria was endemic in Europe until the late 17th century, even during the Little Ice Age[23], and persisted in Russia until well into the 20th century. Moreover the means of combating, if not eradicating, malaria are well established, and not particularly expensive. It is a bitter irony that one of the main reasons why, even today, some two million children in the developing world die every year from malaria, is the malign success of the scaremongering campaign by Western environmentalists to prevent the use of DDT.[24]

Somewhat chastened, the IPCC's 2007 Report confines itself to concluding, with baffling ambiguity, that so far as malaria is concerned, 'Climate change is expected to have some mixed effects, such as the decrease or increase of the range and transmission potential of malaria in Africa,' and then throws into the pot a whole range of other health problems such as malnutrition, diarrhoea, and cardiovascular diseases.[25]

There are, of course, very serious health problems of many kinds throughout much of the developing world which need to be tackled in their own right – global warming or no global warming – much more urgently than they are being tackled at the present time. There is no medical mystery about how to do so. Once this is done, the potential adverse health effects of warming disappear into insignificance. Indeed, the connection is, if anything, the other way round. In the developing world, the major cause of ill health and the deaths it brings, is poverty. Faster economic growth means less poverty but – according to the man-made CO_2 warming theory, incorporated in the IPCC's scenarios – a warmer world.

Warmer but richer is in fact healthier than colder but poorer.

Not that the global warming alarmists confine their health concerns to the developing world. When Sir John Houghton gave evidence to the House of Lords Select Committee he warned us that 'the heat wave in Europe in 2003 … was responsible for the deaths of 20,000 people … The best estimates we have at the moment, if the trend in warming continues, are that that sort of summer will be the average summer in Europe in 2050.'[26] This was, of course, a local rather than a global phenomenon. But it is true that a large number of very elderly people died in Europe that summer, and the highest incidence of death by far was in France, where some 15,000 very elderly people died of dehydration.

As it happens, I spent the summer of 2003 in south-west France myself, and found it perfectly tolerable, but it was clearly a hardship for some. It is the custom in France for every family to go away on holiday during the first fortnight in August, leaving behind, to fend for themselves, those family members who are too old to travel. In August 2003 this proved to be a problem (the number of staff at old people's homes who had gone on holiday at the same time did not help, either). In its wake the Minister of Health saved his skin by sacking the unfortunate senior civil servant who held the post of Director of Public Health and setting up an official enquiry. As a result of the report from that enquiry, arrangements have been put in place (the annually updated *plan canicule*), which – at trivial cost – will prevent a repetition.

So much for yet another scare – not to mention the fact that, in most of Europe, many more very elderly people die of hypothermia in winter than die, or are likely to die, of dehydration in summer.[27] So in Europe, at any rate, warmer weather will save lives. Indeed, in the wake of the 2003 heat wave, a Department of Health study for the UK predicted an increase in annual heat-related mortality of 2,000 and a *decrease* in annual cold-related mortality of 20,000 by the 2050s, using the Hadley Centre climate model.[28]

Indeed, according to the IPCC, this appears to be the pattern for the world as a whole. There is a table in its November 2007 'Synthesis Report' which purports to show 'major projected impacts [of global warming] by sector', ranked from 'virtually certain', through 'very likely', down to 'likely'. So far as health is concerned, the only outcome ranked as 'virtually certain' is 'Reduced human mortality from decreased cold exposure'.[29]

What, then, of the IPCC's overall figure for the likely net cost of a warmer world, working, as they are, on the assumption that no measures are taken to curb carbon dioxide emissions, and carefully examining all the likely adverse consequences, and much less carefully the likely benefits?

It will be recalled that the report's best estimates of the likely warming of the planet over the next hundred years range from a rise of 1.8ºC/3.2ºF to one of 4ºC/7.2ºF above the estimated 1980–1999 average temperature, depending on the emissions scenario (or 'story line') chosen. The report then takes the upper end of the range – a 4ºC/7.2ºF warming – and claims that overall, this would mean a loss, by the end of the 21st century, of anything between 1% and 5% of global gross domestic product. It adds that this is the global average figure, and that developing countries will experience larger percentage losses.[30]

Given that this conclusion derives from the top end of the range, and given that the IPCC insists that all its scenarios are of equal validity, it is clear that, on the basis of the IPCC's own methodology, there may be no net cost at all from global warming over the next hundred years: *it may even be beneficial*. But let us err on the side of caution, and take not only the top end of the IPCC's warming range – a rise of 4ºC/7.2ºF over the next hundred years – but also the top end of its projection of the net damages, a loss of 5% of world GDP.

A loss of 5% of world GDP would undoubtedly be a very significant loss indeed, but to put it in perspective we need to do some simple arithmetic. Heeding the IPCC's very proper warning that the loss will be greater than 5% for the developing countries (and thus less than 5% for the developed world), I shall make the calculations on the assumptions of a 10% loss of GDP in the developing world and a 3% loss in the developed world.

Again, to err on the side of caution, let us look at the gloomiest of the IPCC's six scenarios (that which, the reader will recall, the Stern Review arbitrarily chooses as its business-as-usual base case), even though it is not the scenario which generates the 4ºC/7.2ºF temperature rise, but one of 3.4ºC/6.1ºF. This is the scenario which has the lowest rise in living standards, partly because it has the lowest rate of technological advance, but more particularly because, by a long way, it has the highest projected growth of world population – to more than 15 billion by 2100, or no less than 65% higher than the United Nations' 'medium' population forecast for that year, and almost half as much again as UN's highest projection.[31]

According to this scenario, living standards (measured in the conventional way as gross domestic product per person) would rise, in the absence of global warming, by 1% a year in the developed world, and by 2.3% a year in the developing world (these and subsequent assumptions are taken directly from the IPCC's Special Report on Emissions Scenarios). It can readily be calculated – using, to repeat, a cost of global warming of 3% of GDP in the developed world and as much as 10% in the developing world – that the disaster facing the planet is that our great-grandchildren in the developed world would, in a hundred years time, be only 2.6 times as well off as we are today, instead of 2.7 times, and that their contemporaries in the developing world would be 'only' 8.5 times as well off as people in the developing world are today, instead of 9.5 times as well off.

If we were to take the most optimistic of the IPCC's growth scenarios, which is after all the only scenario where the 'best estimate' is warming of as much as 4ºC/7.2ºF, in which living standards rise by 1.6% a year in the developed world and by 4% a year in the developing world (heroic, but think of China and India), the results are even more startling. To be precise, the great disaster facing the world from the prospect of global warming is that our own great-grandchildren would, instead of being slightly more than 4.8 times as well off as we are, be only some 4.7 times as well off. And as for their contemporaries in the developing world, instead of being 50 times as well off as the population of the developing world is today, they would 'only' be 45 times as well off.

Such are the ravages of global warming. This is the bottom line. This is the existential threat facing the globe. This is the disaster from which we are told we have to save the planet. This is the greatest threat facing the people of the world today. If only it were.

Of course, if it were a real threat, and one that could readily be avoided, it would certainly be worth avoiding it. But even if (and it is a big 'if') global warming could realistically be averted by cutting back drastically on carbon dioxide emissions, that is very far from costless. So we need to reflect long and hard on how big a sacrifice the present generation and their children should be asked to make in order to make it more likely that the generation a hundred years hence, instead of being many times as well off as we are today, will be even better off. And if we are fearful of appearing selfish, we can look at it the other way round. How great a sacrifice do we think the (very much poorer) people of Victorian England should have made to cut back on carbon dioxide emissions (for that was when it all began) at the birth of the industrial revolution?

Indeed, perhaps in hindsight the industrial revolution was all a ghastly mistake, and perhaps at any moment we will find our political leaders, who are as ready to apologize for the errors of their predecessors as they are unready to apologize for their own, apologizing for the industrial revolution, too.

There will no doubt be some who feel that to analyse the alleged threat of global warming in economic terms – that is, in terms of living standards – is profoundly mistaken, not to say immoral (despite the fact, incidentally, that thc Stern Review, for example, is an attempt to do just that). Are not human lives themselves at stake? It is a fair question, but one that is not difficult to answer.

In the first place, natural disasters such as hurricanes, monsoons, droughts, earthquakes, tsunamis, and even pandemics (the vogue word for what used to be known as plagues), have always occurred, and no doubt always will; to attribute them to global warming is not science but political propaganda.

In the second place, where there is a clear link between human life and temperature, such as deaths from either extreme heat or extreme cold, we find that a warmer world would probably, on balance, save lives. Even more important in this context, if the proposed remedy is to attempt to cut back drastically on carbon dioxide emissions by abandoning cheap, carbon-based energy, the cost in terms of slower economic growth would itself cost lives, in terms both of a slower conquest of poverty (despite, incidentally, the global commitment to this) and the reduced resources available for, among other things, the battle against disease.

And in the third place, in no area of public policy do we in practice regard the saving of human life, important though it is, as paramount at all costs, irrespective of all other considerations. If we did, then (for example) we would impose a speed limit on the roads of probably not more than 10 miles an hour.[32]

But leave aside the ethics, which are important, and to which we shall return; what about the political realities? How great a sacrifice (and inconvenience) is it realistic to expect the present generation and their children to bear, in order to benefit future generations, a hundred years hence, who will, whatever happens, be substantially better off than they themselves are today? And even if the present generation in the rich, developed, world are willing to pay the price to help future generations considerably better off than they are, which I am inclined to doubt, is it politically conceivable that the very much poorer people of the developing world will be willing to do so, or even that their governments will ask them to? Self-evidently, it is not.

It is becoming increasingly clear that, even if the science of the conventional wisdom is correct, the policy prescriptions, which we are told flow from it, are mistaken.

Chapter 3

The Importance of Adaptation

Perhaps the single most serious flaw in the IPCC's analysis of the likely impact of global warming is its grudging and inadequate treatment of adaptation, which leads to a systematic exaggeration of the putative cost of global warming – if, indeed, over the next hundred years there is any net cost at all.[1]

The IPCC prefaces its assessment with the statement that 'The magnitude and timing of impacts will vary with the amount and timing of climate change and, in some cases [*sic*], the capacity to adapt.' But adaptation will *always* occur. Leaving aside what might loosely be termed evolutionary adaptation, which is relevant to ecosystems (the second of the IPCC's five headings), the capacity to adapt is arguably the most fundamental characteristic of mankind. We have adapted to different temperatures over the millennia we have been around, and we adapt today to widely different temperatures around the world. And that adaptive capacity is increasing all the time with the development of technology.

To assess the cost of climate change in the absence of adaptation is about as sensible as assessing the risk of catching pneumonia in London on the assumption that we all go out and about in the cold and the rain in our bathing costumes. Yet to a considerable extent, that is just what the IPCC does. The relevant section of its 2007 Report, for example, contains two tables, one of 'Key Impacts' (SPM-1) and the other of what it describes as 'examples of possible impacts of climate change due to changes in extreme weather and climate events, based on projections to the mid to late 21st century' (SPM-2). Of the first, it explicitly declares that 'adaptation to climate change is not included in these estimations'. Of the second it explains that the 'possible impacts ... do not take into account any changes or developments in adaptive capacity'.

This concept of static 'adaptive capacity' is central to its analysis. Thus in its review of the dangers in different parts of the world, it explicitly acknowledges that, in the case of Australia and New Zealand, these will be limited by the fact that 'The region has substantial adaptive capacity due to well-developed economies and scientific and technical capabilities.' Presumably the same applies to Europe and North America, although, curiously, the IPCC does not say so. But it does express concern about the effect of projected warming on the poorer regions of the world, particularly in Africa and parts of Asia, because of their 'low adaptive capacity'.

This somewhat patronizing judgement seems ill-founded for three reasons. First, as we have seen with the IPCC's own economic growth projections, on which its temperature projections rest, the poorer regions are, for the most part, not going to be poor in a hundred years time. Second, for those parts that do remain poor, overseas aid programmes will clearly be focused on improving their adaptive capacity, should the need arise (this is, incidentally, a much more realistic objective for overseas aid than the promotion of economic development). And third, there will almost certainly be substantial technological development over the next hundred years, which will significantly enhance adaptive capacity worldwide, in many cases far beyond what it is at the present time.

In short, the IPCC's analysis and conclusions are seriously undermined by the systematic underestimation of the benefits of adaptation, deriving both from its assumption that 'adaptive capacity' is severely and permanently constrained by economic underdevelopment in the developing world, and its assumption that, for the world as a whole, it is constrained by the limits of existing technology – that is, the assumption that there will be no further technological development over the next hundred years. This last is clearly absurd in the important case of agriculture and food production, and is implausible in general. As a result, the IPCC's overall cost assessment inevitably suffers from a pronounced upward bias.

To be fair to the IPCC, it has to be said that the British government's alarmist Stern Review is even more deficient when it comes to adaptation, as indeed it is on every other aspect of the global warming debate for that matter. A characteristic example of its approach is its stark warning that 'a recent study predicts up to a 70% reduction in crop yields by the end of this century under these [high temperature] conditions, assuming no adaptation'. Not only does the assumption of no adaptation make the estimate completely worthless, but when we come to the nature of this 'recent study', which is revealed only in a discreet footnote in the subsequent chapter, we discover that 'strictly speaking [I like that] these results are for groundnuts only' – and then only in northern India.[2]

So far I have treated adaptation as an integral factor in assessing the true impact of warming, as indeed the IPCC does (albeit inadequately), since a great deal of it will occur autonomously as a result of the innate adaptability of mankind, and the response of the market to changing circumstances. The so-called 'dumb farmer' hypothesis, which characterized the IPCC's previous (2001) report, according to which damage was calculated on the assumption that, as the world got steadily warmer, farmers would carry on growing precisely the same crops, in precisely the same places, in precisely the same way, was clearly a nonsense, and has rightly been abandoned.

But some forms of adaptation, such as the creation or improvement of sea and flood defences, would, if and when they became necessary, require government intervention. The IPCC, needless to say, adopts its characteristically downbeat approach to this, declaring that 'Adaptation for coastal regions will be more challenging in developing countries than developed countries, due to constraints on adaptive capacity.' It must be said that the challenge ought to be a manageable one – the Dutch, after all, managed it pretty effectively even with the technology of the 16th century, and technology has scarcely stood still over the past half-millennium. But this might well be a suitable focus for overseas aid, should the need arise.

Again, history suggests that rational adaptation to the vagaries of the climate may include some degree of population movement, albeit on nothing like the scale of the economic migration that occurs for non-climatic reasons, not to mention the much more brutal movements that wars and other human conflicts have caused in the past, cause today, and may cause in the future. This, too, would be a suitable focus for international aid, should the need arise.

So adaptation can also be seen as a deliberate policy response, and can be compared with the alternative response, embodied in the conventional wisdom, of seeking to control the world's temperature by severely limiting carbon dioxide emissions.

Looked at in this way, the superior cost-effectiveness of adaptation is clear, for at least six reasons:

First, none of the adverse impacts identified by the IPCC are new ones. Drought, hunger and disease long plagued many parts of the developing world before there was any suggestion of man-made global warming. Much the same applies to the flooding of low-lying coastal areas, although this tends to occur (as in Bangladesh) as a result of rivers overflowing their banks in the monsoon season, rather than from sea-level rise.[3] Global warming is essentially projected to exacerbate problems that already exist – and indeed the projected increase in damage is a small fraction of the damage these afflictions already cause. So directly addressing these problems more vigorously will not only bring substantial benefits, but will do so even if there is no further global warming at all.

Second, adaptation will substantially reduce the adverse impact of any future global warming that may occur, even if it turns out that the link between atmospheric greenhouse gas concentrations and global temperature has been greatly exaggerated.

Third, as we have seen in Chapter 1, while warming may be global, any adverse impact is subject to considerable local variation. Adaptation enables us intelligently to tailor our response to this variation.

Fourth, there are benefits, as well as costs, from global warming. Adaptation would enable us to pocket the benefits, while diminishing the costs.

Fifth, the beneficial results of adaptation arise far more quickly than is even theoretically possible from the emissions-cutback route (which incidentally is revealingly known as 'mitigation' – the climate scientists who first chose this term presumably recognizing that 'prevention' would be unjustifiably hubristic, although in plain English it is adaptation which is clearly the means of mitigating the adverse consequences of warming).

And *sixth*, unlike mitigation, which to be effective requires, as we shall see, an enforceable global agreement, which is unlikely to be attainable, adaptation is essentially a matter of a large number of local and practical measures, which require no international treaty or worldwide agreement for their implementation.

In short, even if the conventional scientific wisdom is correct, there remains the fundamental question of what is the most cost-effective way of addressing the likely consequences of global warming. Is it to adapt to them, as man has adapted throughout the ages and throughout the world to the vagaries of the climate, or is it to attempt to prevent them, even if this means radically transforming the global economy?

The adaptation dimension is of crucial importance. Regrettably, the IPCC's belated acceptance of this is distinctly half-hearted. For example, the adaptation and mitigation section of its final Synthesis Report declares that 'financial, technological, behavioural, political, social, institutional and cultural constraints limit both the implementation and effectiveness of adaptation measures', yet it discovers no such constraints in the way of mitigation, which in reality is bound to prove far more costly and technologically challenging. When it comes to mitigation measures, the worst constraint the IPCC is prepared to concede is that 'Resistance by vested interests may make them difficult to implement.'[4]

The perspective of the British Government's Stern Review, with its insistence on the urgent need for drastic action by the governments of the world to curb carbon dioxide emissions, is at least equally cockeyed. Thus even before the review was published, its author declared that 'Adaptation is particularly difficult when the precise nature and incidence of effects are uncertain.'[5] There speaks the true bureaucrat, with no real understanding of the very essence of a competitive market economy, which, as Hayek pointed out, is that it is a discovery process. As he wrote almost half a century ago:

> Competition produces an adaptation to countless circumstances which in their totality are not known and cannot be known to any person or authority … We know the general character of the self-regulating forces of the economy and the general conditions in which these forces will function or not function, but we do not know all the possible circumstances to which they bring about an adaptation. This is impossible because of the general interdependence of all parts of the economic process; that is, because, in order to interfere successfully on any point, we would have to know all the details of the whole economy, not only of our own country but of the world.[6]

Essentially, adaptation will enable us, if and when it is necessary, greatly to reduce the adverse consequences of global warming, at far less cost than that of mitigation, to the point where for the world as a whole, these are unlikely greatly to outweigh (if indeed they outweigh at all) the customarily overlooked benefits of global warming. If on top of this we were, on belt and braces grounds, to embark on a policy of cutting back drastically on global carbon emissions (always assuming that this is politically practicable, which as we shall see in Chapter 5 is highly doubtful), the residual benefit would be marginal and the cost massive.

There are, of course, some for whom the choice between adaptation and mitigation is a matter neither of political practicability, nor even of which is the more cost-effective: rather it is ideological. For them the issue is simple: man has been messing about with the planet in general, and with the climate system in particular, and he must mend his wicked ways and desist. They are clearly entitled to their view, although I doubt if – for good reason – it is very widely shared. I suspect there are few people, when they come to think about it, who regard the huge improvement in living standards, including a substantial reduction in infant mortality and a substantial rise in life expectancy, that cheap, carbon-based energy has made possible, as an unwelcome turn of events.

Moreover, with the best will in the world, it is difficult to see how their view could be put into practice. For a start it would mean unwinding the industrial revolution and returning to a pastoral society. But that would indeed be only a start. For man began messing about with the planet when he ceased being a hunter-gatherer and invented farming. So agriculture, which incidentally is in itself a major source of man-made greenhouse gas emissions, largely, but by no means wholly, in the form of methane, would have to go, too. If, on the other hand, their desire is to unwind material progress a little, but not completely – 'so far, but no further' – then their position is entirely arbitrary, and the ideological purity has disappeared.[7]

There are, however, others who argue not from woolly-minded, if sometimes well-intentioned, green ideology, but who fear that perhaps the threat now facing the planet is so dire that adaptation will not be possible, and that a drastic reduction in carbon dioxide emissions, here and now, is therefore essential. Surely the IPCC's measured prose, and the commendable trouble to which it goes to take a rational, if biased, approach to this difficult issue, must be mistaken. Have we not been warned by Mr Gore's alarmist film and the UK Government's Stern Review, neither of which, needless to say, conducted any original research into the subject, of unimaginable catastrophes if radical action is not immediately undertaken?

Indeed we have. So let us take a look at these alleged catastrophes.

Chapter 4

Apocalypse and Armageddon

The Stern Review is at the extreme end of the alarmist camp, warning us that, in the absence of immediate radical action, 'at some point' we may see 'deaths of hundreds of millions of people ... social upheaval, large scale conflict ... major, irreversible changes to the Earth system ... [which] may take the world past irreversible tipping points'[1]. And there is much more in this vein.[2] The phrase 'at some point', incidentally, is not accidental – the review thinks nothing of looking, not merely a hundred, but two or three hundred, and even, at one point,[3] a thousand years ahead. But this sort of scaremongering is not the language of the scientists – and I refer here, not to the sceptics, but to those mainstream climate scientists who subscribe fully to the anthropogenic global warming theory.

Thus, in the immediate wake of the publication of the Stern Review and the endorsement of its most apocalyptic warnings by Mr Blair and his colleagues, Mike Hulme, Professor of Environmental Sciences at the University of East Anglia, and director of the Tyndall Centre for Climate Change Research, which for years has been urging the government to take more effective measures to curb carbon dioxide emissions, accused the politicians of 'actively ignoring the careful hedging which surrounds science's predictions'. He further pointed out that 'to state that climate change will be "catastrophic" hides a cascade of value-laden assumptions which do not emerge from empirical or theoretical science'. Professor Hulme was particularly scathing about Mr Blair's open letter to EU Heads of State, in which he declared that 'We have a window of only 10–15 years to take the steps we need to avoid crossing a catastrophic tipping point.'[4]

A similar message emerged from a booklet published a few months later[5], the principal author of which was Professor Paul Hardaker, now chief executive of the Royal Meteorological Society, and for many years a senior executive at the Met Office (and, towards the end of his time there, a policy advisor to the government on climate change).

Some extracts give the flavour:

> At the present time we cannot attribute individual extreme weather events to climate change. We should distinguish between the possible effects of *predicted* climate change and the extreme weather that is part of the normal variability of the climate.
>
> We know that the climate has changed abruptly of its own accord before. But the idea of a point of no return, or a 'tipping point', is a misleading way to think about climate and can be unnecessarily alarmist. Although climate and weather are fast-moving fields of science, the view of experts is that the best estimate of global temperature rise is between 2ºC/3.6ºF and 4ºC/7.2ºF by 2100.
>
> There is little evidence that the retreating glaciers can be blamed on rising temperatures, and hence on human activity.
>
> At the moment … a Gulf Stream collapse is seen as very unlikely.
>
> Mankind has never been able to control the weather or climate but has, historically, been able to adapt to changes, surviving ice ages and desertification.

It would be wearisome to examine all the various catastrophes regularly predicted by the climate change alarmists, so I shall confine myself to the three most dramatic: a sharp rise in the intensity of hurricanes, of which hurricane Katrina, which devastated New Orleans in 2005, is said to be the precursor; the melting of the Arctic and Antarctic ice sheets, leading to massive sea-level rise and millions of deaths from coastal flooding and inundation; and a shut-down, or even reversal, of the Atlantic thermohaline circulation, commonly (if inaccurately) known as the Gulf Stream, which, it is claimed, has hitherto kept Europe's temperatures some 8°C/14.4°F warmer than they would otherwise be (so global warming might make Europe seriously colder).

First, hurricanes. In February 2006, in the wake of hurricane Katrina, which had devastated New Orleans the previous year, the World Meteorological Organization held a conference in South Africa, at which an international panel of tropical storm experts reported on an investigation into whether global warming was having an effect on global tropical cyclone activity. In the words of the tropical storm expert from the Met Office, the UK's representative on the panel: 'The main conclusion we came to was that none of these high-impact tropical cyclones could be specifically attributed to global warming.' Adding that 'there is no conclusive evidence that climate change is affecting the frequency of tropical cyclones worldwide', he stated that 'there is ongoing debate as to whether it is affecting their intensity'.

The relative intensity of different 'high-impact tropical cyclones', popularly known as hurricanes, over the past century, is difficult to determine, since prior to 1970 we had no reliable estimates of hurricane wind speeds. But we do have reasonably reliable estimates of the damage each have caused, and these have been carefully 'normalized' to take account not only inflation, but also the growth of population, wealth, and in particular, property values in the affected areas.

On this measure, which is the best available, of the ten most severe Atlantic hurricanes since 1900, five occurred in the first half of the period and five in the second half. Seven out of the ten occurred before 1975 – that is to say, before the period when the bulk of the modest 20th century global warming began. The worst of all, by far, was the Great Miami Hurricane of 1926.[6] Katrina, in 2005, was the second worst, and led to confident predictions that 2006 would be another terrible year. In fact, 2006 proved to be among the quietest of the past 20 years. It was to a considerable extent as a result of this that the hedge fund Amaranth, which believed the predictions and bet heavily on a sharp rise in oil and gas prices from the consequent damage to oil installations in the Gulf of Mexico, went bust in one of the most spectacular financial collapses of 2006.

In the eyes of the insurance industry, there has of course been a significant rise in hurricane damage over the years, but that is simply because of the rise in population and property values in the affected areas, which has inevitably caused an increase in tropical storm damage costs, well above the rate of inflation. In the words of the World Meteorological Organization:

> Though there is evidence both for and against the existence of a detectable anthropogenic signal in the tropical cyclone climate record to date [this was written in late 2006], no firm conclusion can be made on this point ... No individual tropical cyclone can be directly attributed to climate change ... The recent increase in societal impact from tropical cyclones has largely been caused by rising concentrations of population and infrastructure in coastal regions.[7]

Next, the melting of the ice sheets. Clearly, the melting of floating polar ice cannot cause any rise in sea levels – just as the melting of ice cubes in your glass of water cannot cause the water to overflow the glass. The issue is solely about the land-borne ice at the poles.

So far as the Greenland ice sheet is concerned, there is no

evidence that melting, or rather, net ice loss, is occurring to any significant extent. This is perhaps not particularly surprising; not only is Greenland a pretty cold place, but there is no evidence at the present time that it is getting any warmer. The most recent study of Greenland temperature records over the past hundred years and more[8] found that the warmest decades so far were the 1930s and 1940s, with 1941 the warmest year of all. By contrast, the last two decades of the century, the 1980s and 1990s, were colder than any of the previous six decades. While there has been some slight warming this century, Greenland temperatures are still below the levels of the 1930s and 1940s. It is evident that the climate in Greenland fluctuates unpredictably for reasons that appear to have little or nothing to do with the greenhouse effect.

But the overwhelming mass of polar land-borne ice (and thus the most significant land-borne ice in terms of its potential effect on global sea levels) is not over Greenland in the north, but over the vast continent of Antarctica in the south. Here it is perfectly true that the West Antarctic ice sheet which covers the peninsular which points its finger towards the southern tip of South America, is showing evidence of melting and glacier retreat. But the West Antarctic peninsular accounts for only around 10% of Antarctic land-borne ice, and has a different climate from the rest of Antarctica. In most of the other 90% of the continent, according to the most recent research, the ice sheet appears to be growing.[9] A further complication is that any warming over the Antarctic is likely to lead to greater snowfall, leading in turn to a thickening of the Antarctic ice sheet.

In the light of the evidence it is scarcely surprising that the worst the IPCC Report can come up with, so far as the polar ice sheets are concerned, is the conclusion that:

> There is medium confidence that at least partial deglaciation of the Greenland ice sheet, and possibly the West Antarctic ice sheet, would occur over a period of time

> ranging from centuries to millennia for a global average temperature increase of 1–4°C/1.8–7.2°F.[10]

The idea that anything sensible can be said about the likely state of the world thousands of years ahead, still less that we can take rational policy decisions on this basis, is mind-boggling. But while we are peering this far ahead, it is worth recalling that even the Hadley Centre concedes that, over a timescale of millennia, the earth may well enter another ice age, a far more damaging prospect than the consequences of warming.[11] There are, in fact, a number of reputable climate scientists who believe – as indeed Professor Lovelock and others did in the early 1970s – that the next ice age may be upon us much sooner than that. But while clearly not impossible, there is so far no evidence to suggest this.

Finally, let us take a look at the Gulf Stream, more properly the Atlantic thermohaline circulation, and the meridional overturning circulation (MOC) of which it is a part. Although there is ample evidence of fluctuations in the strength of these currents from time to time, research has shown no sign of any secular slowdown over the past decade.[12] Nor is there any reason to suppose that there will be, even if there is further global warming over the coming decades. As the eminent oceanographer Professor Carl Wunsch has pointed out, the Gulf Stream is largely a surface current, and thus a wind-driven phenomenon, and 'As long as the sun heats the Earth and the Earth spins, so that we have winds, there will be a Gulf Stream.'[13]

It is not at all clear from its report that the IPCC is fully aware of this, but whether it is or not, the worst it can manage to say about all this is that: 'It is very unlikely that the MOC will undergo a large abrupt transition during the 21st century. Longer-term [millennia again?] changes in the MOC cannot be assessed with confidence.'[14]

In any event, it is now widely believed that the Gulf Stream plays only a minor part in bringing about the mildness of the western European climate, and that a more important factor is what the scientists call 'atmospheric heat transport', or in plain language, warm air currents (from the prevailing south-westerly wind).[15]

It is clear, therefore, that even after looking carefully at the worst nightmare scenarios the alarmists can conjure up, there is no reason to qualify the conclusion reached in Chapter 2, that, on the most pessimistic assumptions, and on the basis of the majority view of the science of global warming, the existential threat confronting the world today, from which the planet must be saved, is that, a hundred years from now, the people of the developing world may not be 9.5 times as well off as they are today, but only 8.5 times as well off.

So the overriding policy question remains: how great a sacrifice is it either reasonable or realistic to ask the present generation – particularly the present generation in the developing world – to make in the hope of avoiding this outcome?

Chapter 5

A Global Agreement?

To the extent that there is a problem of global warming, it is manifestly a global problem. And if the chosen policy for addressing it is to cut back on carbon dioxide emissions, the cutback clearly has to be global, too. Thus the perspective of the developing world is of the first importance. And it is in the developing world, particularly China and India, where emissions are growing fastest. Indeed, China is very soon set to overtake the United States as the single biggest source of emissions, if it has not done so already, chiefly because its rapidly growing economy is so heavily dependent on energy- intensive manufacturing industry.

Both China and India have made their positions abundantly clear; and it has to be said that they are thoroughly understandable, and reflect the perspective of most of the developing world. The overriding priority of both is to continue along the path of rapid economic growth and development. Only in this way can the widespread poverty which afflicts their people, be relieved. Both observe that the industrialized countries of the western world achieved their prosperity thanks to cheap carbon-based energy, and they believe that it is now their turn to do the same.

They add that if there is now a problem of excessive carbon dioxide concentrations in the earth's atmosphere, it is the responsibility of those who overwhelmingly caused it to remedy it. At the very most, they are prepared to concede that, if and when their emissions per head of population have risen to the levels of emissions per head in the rich world, there might be the basis for an international agreement which would be fair for all. But until then, and for very cogent reasons, there can be no question of their agreeing to any binding restrictions on their emissions, irrespective of whether there is any change in US policy on this issue after the Bush Presidency has come to an end.

It is perfectly true that China has set itself an ambitious target of reducing its energy intensity by 20% by 2010 (compared with 2005), as part of its overall drive to improve the productivity of its economy and to cut costs. But this may well not be achieved, and even if it is, such is China's rate of economic growth that it will not prevent her emissions from continuing to rise at a substantial rate, with each year's annual increase far exceeding the UK's total annual emissions.

The current Chinese five-year plan does indeed also include a commitment to reduce the emission of air pollutants, which are a major problem in China. But what is meant by air pollutants are sulphates and other noxious aerosols, not CO_2, which, as we have seen, is not a pollutant. So far as carbon dioxide emissions are concerned, the relevant fact is that China is in the midst of a sustained, multi-year programme of building a new, large, coal-fired power station every five days, increasing its power-generating capacity each year by roughly the equivalent of Britain's total capacity. China is already the world's largest coal producer, yet despite that, and despite its massive indigenous reserves, its coal consumption is growing so fast that it has now (in 2007) become for the first time, a net importer of coal.

What China does not officially admit is that its desire to go flat out for the fastest rate of economic growth it can manage, is not entirely due to its leaders' humanitarian concern for the condition of its people. They also fear that without it there may be serious social unrest and political upheaval.

China also makes two other points, both of which are in a sense merely debating points, but the second of which, as we shall see, has important practical implications. Both these points were voiced at the time of the G8 Summit in Heiligendamm, Germany, in June 2007, at which the death knell of the so-called Kyoto approach to international agreement on this issue was effectively sounded, to be confirmed six months later at Bali.

The first was an observation by Mr Ma Kai, the head of the Chinese National Development and Reform Commission. Warning, reasonably enough, that 'The consequences of inhibiting their [the developing countries'] development would be far greater than not doing anything to fight climate change,' he added that 'Without China's strict family-planning policies, the country's population would have increased by 138 million since 1979, resulting in an extra 330 million tonnes in emissions.'[1]

The second argument was advanced by China's Foreign Ministry spokesman Mr Qin Gang: 'China is now a world factory. The developed countries moved a lot of manufacturing industries to China. A lot of the things you wear and you eat are produced in China. On the one hand you increase the production in China, and on the other hand you criticize China on the emission reduction issue, so it is unfair.'[2] But while to some extent this is indeed a debating point, since China has been more than happy to assume the role of the workshop of the modern world, it is much more significant than that.

As an authoritative recent study has pointed out,[3] the UK has managed to limit its recorded growth in carbon dioxide emissions at modest cost, only by effectively outsourcing them to the developing world, in particular to China. Not only is this futile in the context of the conventional global warming wisdom, but the cost to the UK, and indeed to the European Union as a whole, of the very tough emissions targets which have now been agreed, not to mention the even tougher ones with which the UK has decided to saddle itself unilaterally, are likely to be very much greater. This is not only because the targets themselves are so tough, but also because the scope for the further outsourcing of emissions, although far from negligible (the UK manufacturing sector could well dwindle still further), is plainly less than before. The same point applies, to a greater or lesser extent, to other developed countries.

The authors of the study conclude that 'if the UK wishes to back its leadership claim in the global climate change debate', its target should be in terms not of greenhouse gas *production*, but of greenhouse gas *consumption*. Now that really would set the cat among the pigeons.

Be that as it may, China has made its position clear. As its President, Hu Jintao, declared to the G8:

> Considering both historical responsibility and current capability, developed countries should take the lead in reducing carbon emissions and help developing countries to ease and adapt to climate change.[4]

In other words, while China is adamant that it will not be part of any international agreement that obliges it to cut its emissions, it is prepared to accept both mitigation and adaptation aid from the West. And this is more or less endorsed by the Stern Review, when it declares that 'Securing broad-based and sustained international co-operation to tackle climate change depends upon finding an approach widely understood as equitable ... Given the ability to bear costs and historic responsibility for the stock of GHGs, equity requires that rich countries pay a greater share of the costs.'[5]

Fair enough. No doubt China would, for example, be prepared to install, at heavy cost, carbon capture and storage facilities at all its coal-fired power plants, if and when the technology at some future date becomes a practical proposition, provided that the cost were borne by its Western competitors. But the notion that the peoples of Europe, or the United States, or Japan would be prepared to pay this price is less than compelling. Indeed, at the present time it is hard enough to hold the line against the imposition of barriers to Chinese competition.

The position of India, which like China has vast indigenous reserves of cheap coal on the back of which it is powering its economic growth, is essentially the same. Moreover, so far from being ready to sign up to deliberately making carbon-based energy more expensive, India, like China, massively subsidizes the cost of energy, a practice neither country has any present intention of abandoning. It has been estimated that China's energy subsidies (notably, but by no means only, for oil used by farmers) amount to some 1.5% of total government spending, while India's are estimated to be even greater: perhaps some 2% of GDP.

Before the 2007 G8 Summit, the Indian Environment Minister, Mr Pradipto Ghosh, made it clear that India, too, would not accept CO_2 emissions limits that would retard economic growth and damage India's attempts to eradicate poverty. Indeed, following the Heiligendamm summit, the official German news agency reported that 'Chinese President Hu Jintao and Indian Prime Minister Manmohan Singh have created a new alliance to spearhead emerging economies' opposition to developed nations seeking to impose limits on their greenhouse gas emissions.'[6]

So where does this leave the prospect of an effective global agreement to prevent the further growth of carbon dioxide concentrations in the atmosphere? Not, it has to be said, in very good shape.

It will be recalled that the attempt to forge a global agreement began with the holding of the United Nations Framework Convention on Climate Change in 1992. This led to the signing in 1997, during the Clinton Presidency in the United States, of the Kyoto Protocol, according to which the countries of the developed world agreed, subject to ratification, to reduce their emissions, by 2012 at the latest, to a level some 5% below 1990 levels. It was calculated at the time that if every signatory ratified Kyoto and subsequently met its emissions target, the world's temperature by 2100 would be 0.1°C/0.2°F less than would otherwise be the case – a trivial amount.

The Kyoto Protocol had been explicitly modelled on the Montreal Protocol of 1987, under which the developed nations agreed to phase out the production and use of chlorofluorocarbons (CFCs) after it had been discovered that these were causing a potentially dangerous depletion of the earth's ozone layer. But it was never a realistic precedent. Not only was there a much greater degree of scientific consensus on the effect of CFCs, but the Montreal agreement, unlike Kyoto, covered only a small number of chemicals of marginal economic significance.

Despite this, it was hoped by its supporters that Kyoto, however modest in its ambitions, would be the first step on the road to a much more substantial agreement for the period after 2012. But President Clinton made no attempt to persuade the US Congress to ratify it, recognizing that this would be politically impossible (the Senate had fired an eloquent warning shot across his bows by voting 95 to zero against ratifying any treaty of the kind Kyoto turned out to be). Clinton's successor, President George W Bush, formally announced in 2001 that the US would definitely not be ratifying Kyoto. The principal (although not the only) American objection was, and remains, the fact that the rapidly growing developing countries, notably China, India and Brazil, are effectively outside the process and are determined to remain so.

Now, two-thirds of the way through the Kyoto period (1997–2012), it is pretty clear, with only one exception, that not even the Kyoto ratifiers – neither the European Union, Canada, nor Japan – are likely to meet their modest Kyoto targets. The exception is Russia, where the collapse of the Soviet Union and rapid contraction of its massive, energy-intensive defence sector, which had been such a drain on the Soviet economy, has inevitably reduced emissions substantially. In fact, since 1997, when Kyoto was agreed, CO_2 emissions have grown more rapidly in Canada, and only slightly less so in Europe, than they have in the non-ratifying United States.[7]

So far as Europe is concerned, the EU's 2012 overall Kyoto emissions target is a reduction of 8% below 1990 levels, with the 15 individual member states being assigned national targets to achieve this. Only two of the 15 seem likely to do so: Sweden, which has become increasingly post-industrial, and the UK, where the privatization of electricity generation led power stations to switch on a large scale from coal to gas, which was then much cheaper (the old state-owned gas industry had refused to supply the power stations with gas, which generates roughly half the CO_2 emissions per unit of electricity produced that coal does).

It was against this unpromising background that the German Chancellor, Angela Merkel, with the enthusiastic support of the UK, sought to persuade the G8 to an even more ambitious target for the period beyond 2012. She proposed a firm commitment to cutting emissions by 50% by 2050, and a limit in the world's temperature rise to 2°C/3.6°F. Quite apart from the problem of other countries' emissions, and the fact that those of China and India had almost doubled since 1990 and continued to roar ahead, it was unclear how Chancellor Merkel and her supporters planned to control the sun.

The idea was to isolate the United States and oblige it to give way. The Europeans hoped that President Bush would come under domestic pressure to do so from the Democratic-led Congress. He did not (not even from Mr Gore) and in the event it was Europe that found itself isolated. A US–Japanese counter-proposal, warmly supported by both Russia and Canada, rejected any notion of mandatory or unilateral G8 emissions limits. It successfully suggested instead that the G8 should agree to 'seriously consider' halving emissions by 2050, providing the developing countries, in particular China and India, agreed to participate in the process. As we have seen, the developing countries have made clear they will not agree to participate – at least not in a way acceptable to the United States and its supporters.

It is perfectly true that spokesmen for both the United States and the major developing countries are from time to time prepared to pay lip service to the idea of a global agreement on limiting emissions, provided the burden of doing so is equitably shared. But what the United States considers an equitable sharing of the burden is worlds apart from what China and India consider equitable, and there is no prospect whatever of this chasm – it is far more than a gap – being closed.

This became even more apparent six months later when the circus moved on, greatly enlarged, from the ancient Baltic seaside resort of Heiligendamm, to the rather more exotic Indonesian seaside resort of Bali. The occasion was the global conference, under the auspices of the United Nations, summoned to pave the way to a successor to the Kyoto accord, which expires in 2012. The European Union, with Germany and the UK at the fore, came determined to secure agreement that the developed world should accept a mandatory global emissions reduction of between 25 and 40% by 2020.

Their failure was total. The United States on the one hand, and China and India on the other, held fast to the positions they had staked out at the G8 summit six months earlier, and the rest of the developed world – including Japan, Russia, Canada and even Australia, whose newly-elected Labour government had made ratification of Kyoto its first act on taking office (an empty gesture if ever there was one) – were in effect content to hide under America's skirts. Once again Europe was isolated, and, to the dismay of the world's Greens, the fortnight-long meeting ended with no targets of any kind for emissions reductions, merely a commitment to further talks, euphemistically known as the 'road-map', and the hope, which there is little evidence to sustain, that the US will undergo a

Damascene conversion on radical mandatory emissions reductions once the Bush era has come to an end.[8]

This, then, is where we are now. No doubt there will in due course be international agreement on significant adaptation aid for the poorer countries, and quite possibly on non-carbon technology transfer from the developed to the developing world, although this last was sought in vain at Bali. But what is abundantly clear is that the Kyoto approach is dead and buried. Admittedly the European Union is still, theoretically, committed to going it alone, having agreed in principle to cut its emissions by 20% (below 1990 levels) by 2020, with the UK going even further, introducing a Climate Change Bill which imposes, unilaterally, a statutory reduction of 60% in CO_2 emissions by 2050. Indeed, at the time of writing there is talk of the government amending the Bill to raise the statutory requirement to an 80% reduction.

But it is difficult to see the point of this costly masochism. The thinking behind it (if it can be dignified by the name) appears to be that we shall be giving a moral lead to the world, which the world will then follow. It is essentially the same approach the CND used to take – that the moral force of the UK renouncing nuclear weapons would persuade other nuclear powers to do the same. It was not convincing then and it is even less convincing now – the UK was, and to some extent still is, a significant nuclear power: whereas it accounts for less than 2% of global CO_2 emissions. Nor is it comforting that this is leadership akin to that of the Earl of Cardigan at the charge of the Light Brigade.

The European Union also agreed, in March 2007, that a 'binding' target of 20% of its energy consumption should come from renewable sources by 2020. It is quite clear that Mr Blair enthusiastically signed up to this without having the slightest idea of what it meant. An internal Whitehall departmental study, subsequently leaked to the press, spells it out.[9] In a nutshell, the 19-page study concludes that to meet this renewablestarget would cost the UK, by 2020, between £18- and £22 billion a year in increased electricity costs alone, that this is three times the cost of attempting to achieve the same amount of emissions reduction by taking full advantage of the EU Emissions Trading Scheme (about which more in the next chapter), and that the target completely 'lacks credibility'.

In any event, the problem with one or more countries going it alone is not simply the futility of the moral leadership conceit, nor even the heavy cost to those who do so (although we will need to take a good look at this in the next chapter), it is also the nugatory reduction in overall global emissions that this would lead to.

This is because the only practical way of cutting back on carbon dioxide emissions is to raise the cost of carbon-based energy, so that energy-saving becomes more attractive, and non-carbon-based energy more competitive. But as energy prices in Europe rise, with the prospect of further rises to come, energy-intensive industries and processes would progressively decline in Europe, and expand in countries like China, where cheap energy would continue to be available (a process known to the IPCC as 'leakage'). No doubt Europe could, at some cost, adjust to this, just as it has to the migration of most of its textile industry to China and elsewhere, thanks to their cheap labour. But it is difficult to see the point of it. For if carbon dioxide emissions in Europe are reduced, only to see them further increased in, for example, China, there will be little if any net reduction in global emissions at all.

Not that Europe's policy is particularly consistent in any event. Under the European Union's competition rules, state aid is, in general, illegal. But there are a few exceptions, perhaps the most important of which is state aid to the coal industry, currently running at some 5 billion euros a year – more than half of it in Germany. (Most of the rest is in France and Poland. There is none in the UK, thanks to the phasing out of coal subsidies, which began during the Thatcher government of the

1980s – not because we were opposed to coal, but because we were opposed to subsidies.) In May 2007, only a month before the G8 summit, the European Commission decided that this waiver should continue unchanged. Germany also subsidizes new coal-fired power plants by granting them extra permits under the European Union's emissions trading scheme (we shall be taking a look at that in the next chapter). And the German electricity industry is currently embarking on a multi-billion euro programme of renewing its ageing power stations, mainly with new coal-fired units.[10]

All this underlines the fact that, even if there were to be a global agreement on emissions limits to make sense of this approach (which there will not be), cutting back on carbon dioxide emissions is far from easy. The principal reason for this is the cost of doing so. It is to that which we now turn.

Chapter 6

The Cost of Mitigation

'Mitigation', as we have seen, is the term customarily used to describe the attempt to prevent further global warming by cutting back carbon dioxide emissions sufficiently to stabilize CO_2 concentrations in the atmosphere. Quite apart from the fact that stabilizing carbon dioxide concentrations is not the same as stabilizing the global temperature, it is important to be clear that what is implied by this approach is a very severe cutback indeed. According to the Hadley Centre, 'Only by a reduction of about 70% in [global] carbon dioxide emissions would we be able to stabilize its concentrations in the atmosphere,' but it adds that feedbacks between climate and the carbon cycle 'may mean that the reduction in emissions required to stabilize carbon dioxide concentrations in the atmosphere would be even larger than 70%'.[1]

Three conclusions immediately follow from this. The first is that, given the continuing brisk growth in global emissions since this was written, the cutback required, on this analysis, would be even greater today. The second is that a cutback of this kind is even theoretically possible only if there is a global agreement in which the developing countries participate to a significant extent, which (as we have seen) is not on offer – at least for the foreseeable future. And the third conclusion is that, even if this were attainable, it would require drastic changes in the way we produce and consume energy right across the board – from electricity generation to transport, both surface and air – and in the amount we consume.

Feelgood measures in the western world, from driving a hybrid car to the abolition of plastic bags, to not leaving our television sets on standby, are trivial to the point of irrelevance in this context. Some of them may be desirable for other reasons, while others are distinctly undesirable: to abandon our dishwashers in favour of hand-washing, for example, would simply expose our children and grandchildren to a greatly increased risk of food poisoning, since dangerous bacteria which can survive the temperatures our hands can tolerate cannot survive the very much higher temperatures reached by dishwashers. But what all such do-it-yourself feelgood measures have in common is their complete irrelevance to the scale of the carbon dioxide cutback we are told is required.

So how much would it cost to reduce carbon dioxide emissions per unit of output to the extent allegedly required by switching to non-carbon or low-carbon sources of energy (including adding the process of carbon capture and storage to electricity generation in coal-fired power stations), and by using less of it?

The only honest answer is that we do not know, but all the signs are that it would prove very costly indeed. A key test is to consider how high a carbon tax would need to be in order to generate the necessary change in behaviour, both on the supply side and the demand side. And it is significant that this is something which those politicians who identify global warming as the greatest threat facing the planet, are conspicuously reluctant to discuss, let alone to propose.

Yet as the noted energy economist Dr Dieter Helm told the House of Lords Committee, 'If it turns out that you need a very high carbon tax to get a change of behaviour, it is not possible to argue at the same time that the costs of achieving a carbon reduction are very low.'[2] Another indication of this is that the annual cost to the British taxpayer and energy consumer of support for renewable energy of one kind or another is already running at the best part of a massive £1 billion a year (the government is notably evasive about the precise amount), to meet less than 2% of UK energy needs.

There are basically three ways in which, combined, it is suggested the world economy might become very significantly decarbonized. All of them would require a substantial rise in the price of carbon, in all its forms.

The first is through a reduction in energy intensity. This is indeed occurring all the time, albeit not in a straight line, along with other dimensions of cost reduction per unit of output (or productivity): it is what is commonly known as economic progress. It has been taking place throughout the 20th century warming, and a continuing reduction is assumed in the International Energy Agency's forecast that global energy demand will still be more than half as much again as it is today, by 2030.[3] No doubt a sharp worldwide rise in the price of energy, whether policy-induced or not, would accelerate the trend, although it would require a major technological breakthrough for this to transform the picture.[4]

But even if this were to occur – and indeed to some extent it appears to be built into the IEA's projections, which assume an otherwise implausible marked fall in the rate of growth of energy demand after 2015 – the evidence suggests that what economists call the price-elasticity of demand for energy is not all that great, particularly so far as transport is concerned. The price of oil has almost quintupled over the past six years, and the price of petrol at the pumps has risen substantially as a result, but this has had little effect on the amount we drive or on the speed at which we drive (despite the fact that slower speeds are significantly more economical).

It seems likely that the rise in the price of petrol needed to persuade us to use less, and the rise in the cost of air fares (from a higher aviation fuel price) needed to persuade us to fly less, would have to be very large indeed; and the political cost of this would almost certainly be prohibitive. It is striking (not to say comical) that when, in January 2008, the oil price reached $100 a barrel and UK energy companies announced new and increased tariffs for 2008, Britain's Chancellor of the

Exchequer lost no time in calling for urgent talks on whether these price rises were justified – despite the fact that the British government's climate change policy requires energy prices far in excess of anything so far seen.

Moreover, once again we must not forget what is likely to happen in the developing world, where China and India, for example, are only at the very early stages of the progress to mass car ownership. The plaintive hope of the IPCC that they will avoid it by 'investment in attractive public transport facilities and non-motorized forms of transport'[5] (whatever the latter may mean) is not an expectation shared by those who know these countries.

The third road to substantial decarbonization is a switch to non-carbon-based energy; either nuclear power or various forms of renewable energy. So far as transport is concerned, the front-runner here is biofuels, and in particular ethanol, a fuel produced from crops; usually corn or sugarcane. This is particularly popular among US farmers, who receive a handy government subsidy for producing it, currently running at some $7 billion a year.[6] They are also heavily protected from foreign (chiefly Brazilian) competition, which is among the issues which led to the breakdown of the Doha round of trade liberalization in 2006. And biofuels have now become one of the European Union's latest fads.[7]

But biofuels, such as ethanol, have their downsides. In the first place, as studies have shown,[8] it is far from clear that ethanol produces significantly more energy than is used in its own production. In the second place, it requires a vast amount of land to produce a relatively small amount of ethanol. This not only antagonizes environmentalists, upset by the destruction of rainforests for this purpose, but has also led to a marked rise in food prices, in particular the price of grain. Indeed in June 2007 the Chinese government suspended its production of ethanol explicitly for this reason.

As a recent OECD report points out: 'the potential for the current technologies of choice – ethanol and biodiesel – to deliver a major contribution to the energy demands of the transport sector without compromising food prices and the environment is very limited.'[9] Meanwhile, in Latin America and Asia, deforestation gathers momentum as land is cleared to grow crops to supply the biofuel market. It is clear that biofuels are far more expensive, both economically and environmentally, than their advocates are prepared to admit.

So far as the production of electricity is concerned, which is still by far the largest (although not the fastest growing) user of carbon-based energy, the front-runner, at least in the eyes of the British government, is wind power. This is the principal means by which the government plans to achieve its target of 20% of electricity generated from renewable sources by 2020[10] (not that even its own officials believe that there is much chance of this being closely approached, let alone met). We have already seen the consternation caused in Whitehall when this absurd unilateral aspiration became part of an allegedly binding EU agreement.

Whether forests of wind turbines are a vision of beauty or an environmental outrage is a matter of personal opinion, but what is undoubtedly the case is that wind power, even with the help of the substantial subsidies it currently enjoys, is nowhere near competitive with conventional power stations. The key question is not about the economics of generating electricity from a single windmill, but about the economics of a significantly wind-based electricity supply system.

In the first place this would require heavy additional transmission and distribution costs, which the government largely ignores. Second, and even more important, wind power is intermittent, whereas electricity has to be permanently on tap. Since the cost of storing electricity is prohibitive, this means that conventional (carbon-based) power stations have to be kept in instant readiness as a back-up system for when the wind stops blowing, as it occasionally does, often in very cold weather, thus massively increasing the overall systemic cost of wind power.[11] This, again, the government largely ignores.

A non-carbon-based source of electricity generation that is considerably less uneconomic than wind power is nuclear power. Its use is limited in a different sense, in that it can only be economic as a steady state, base load provider. It cannot be turned on and off to meet the inevitable diurnal and seasonal fluctuations in demand – for that, conventional carbon-based power stations are required. But the UK government's belated decision that 'the electricity industry should, from now on, be allowed to build and operate new nuclear power stations, subject to meeting the normal planning and regulatory requirements' is welcome, if long overdue.[12]

Nuclear power does, however, have hidden costs in the shape of both the storage and disposal of nuclear waste (hitherto, as a result of government dithering, an unresolved problem in the UK) and of safe decommissioning once the power station comes to the end of its life. Nonetheless, it does not require an inordinately large increase in the cost of carbon to make nuclear power fully economic, which is a great deal more than can be said for renewable energy.

However, there still remains the political problem of the widespread public hostility to nuclear power, which is as often as not fomented by those who profess the greatest concern about man-made global warming.

I have no doubt that nuclear power stations can be – and of course must be – made safe, and indeed are safe, although in this age of terrorism, additional precautions are needed. But public apprehension, and the extreme opposition of a vociferous minority, create political problems and make the holding of a public inquiry before a nuclear power station is constructed something of a nightmare. When I was UK Energy Secretary in the early 1980s I set up the planning inquiry into Sizewell B, the last nuclear power station to have been built in Britain. It turned out to be the longest planning inquiry in British history. The next generation of nuclear power stations may well be more economic than their predecessors, but the political problems remain, and it is clearly a matter of decades, rather than years, at best, before they will be able to play any significant role on the energy scene.

Finally, the fourth hoped-for road to decarbonization, is the technology known as carbon capture and storage (CCS), or carbon sequestration, by which carbon-based fuels – coal, oil or gas – are used to generate the electricity, but the carbon dioxide produced in the process is injected underground and thus prevented from escaping into the atmosphere. Given the inescapable practical fact that we are going to be relying overwhelmingly on carbon-based fuels for the foreseeable future (the IEA reckons that even by 2030, coal, oil and gas will be satisfying more than 80% of the world's energy demands, with nuclear, hydropower, biomass and other renewables together accounting for less than 20%), this is potentially the most important route of all.[13]

It is, incidentally, frequently assumed that renewable, non-carbon-based energy, is 'clean' energy. In fact, by far the largest source of renewable energy consumed in the world today, much larger indeed than all other renewable sources taken together, is biomass – essentially, the burning of animal dung, which is still widely used in those parts of the developing world where there is no supply of electricity. As a result, the problem of indoor pollution in the developing world, so far as health is concerned, is probably second only to the lack of clean drinking water and sanitation, causing (it is conservatively reckoned) at least a million deaths a year. The African peasant, desperately seeking to replace his renewable dung with an electricity supply, may not be amused to be told that, if – as is almost bound to be the case – it is produced by a carbon-fired power station, the electricity generated is dirty, not to say polluting, and should be discouraged.

But to return from the reality of the developing world to the wonders of CCS, the problem is that, despite a great deal of research, the technology is still not proven. According to Mr

Jeroen van de Veer, chief executive of the Royal Dutch Shell group, which is particularly active in this field, it will take a decade to test the technology in pilot projects before there can be any question of moving to larger-scale projects.[14] And even if the technology does pass the test, it is likely to require a very high price of carbon, and thus of electricity, to make it economic – not something that is likely to commend it to the poorer countries of the developing world, unless of course they are bribed to install it.

It is sometimes suggested that an additional reason for moving away from carbon-based fuels is concern over energy security. But this argument does not stand up. The instability of the Middle East and the unreliability of Mr Putin's Russia do indeed, at first sight, raise question marks over our supplies of oil and gas, but there is no need to exaggerate the danger. As Adam Smith wisely pointed out more than 200 years ago, 'It is not from the benevolence of the butcher, the brewer, or the baker, that we expect our dinner, but from their regard to their own interest.'[15] Moreover there are many accessible sources of oil around the world, although given Europe's large and growing dependence on piped Russian gas, it might well be prudent for the countries of Europe to enlarge substantially their strategic gas storage capacities, to remove from Mr Putin the temptation to use a temporary, but highly disruptive interruption of gas supplies to the West, as a geopolitical bargaining counter.

Indeed, there has been interminable – and so far inconclusive – discussion about this within the European Union, as there has been about other aspects of Europe's gas industry, in the context of attempting to hammer out an agreed European gas policy. So far Russia has been able to divide and rule, concluding separate gas supply deals with individual member countries, notably Germany. It is time for the UK, at any rate, to cease waiting for a European agreement which seems as far away as ever, and to beef up its own national gas storage capacity, which at present is dangerously small.[16]

The main reason, however, why the needs of energy security do not dictate any move away from carbon-based fuel is coal, of which pretty well all the major energy-consuming nations of the world (including not least the UK) have centuries-worth of indigenous supplies. It is true that, as I well recall from my time as her Energy Secretary, Margaret Thatcher was a deeply committed believer in nuclear power, largely on energy security grounds. But her rejection of coal in this context had nothing to do with carbon dioxide emissions (although she understandably found it a useful debating point) and everything to do with her well-founded distrust of the politically-motivated leadership of the National Union of Mineworkers at that time. That problem is happily long gone.[17]

As for China, its decision to rely heavily on coal rather than imported gas, even though burning coal produces roughly twice the CO_2 emissions that gas does per unit of electricity generated (China also has a significant nuclear power programme, but that is on a smaller scale), is explicitly predicated on considerations of energy security. Much the same is true of India.

For the United Kingdom, there is another and more urgent dimension to the energy security equation. With existing nuclear power stations coming to the end of their lives well before they can be replaced by the next generation of nuclear power stations, the lights are going to go out, despite all the government's windmills, unless new carbon-based power stations are allowed to go ahead, and the lives of existing ones (including coal-fired ones) extended, CO_2 emissions or no CO_2 emissions.[18]

But if western governments, at least in Europe, are going to remain obsessed with the alleged need to cut back sharply on CO_2 emissions by increasing the cost of carbon, what is the best means of achieving this?

The route the politicians (but very few economists) prefer, is known as 'cap and trade', a régime in which emissions (or some of them) are statutorily capped, and the emitters are then free to trade the permits to emit which result from this system. The trouble is that both in theory and in practice (for it has already, to some extent, been put into practice) it is a method that has little to commend it.

For one thing, it is in no sense the 'market' solution that it purports to be. It is essentially a government-controlled, administrative rationing system, in which the rations can subsequently be traded. It is rather as if, instead of seeking to cut back on smoking by taxing it, we were to allocate soviet-style production permits to the cigarette manufacturers, which they were then permitted to buy and sell among themselves. Of course, for the market-makers and other middlemen who trade in the CO_2 emissions permits, it is indeed a market, and one which they will not hear a word said against; for them it presents a lucrative and – they hope – growing business opportunity.

Among its many other drawbacks, cap-and-trade is arbitrary and distortionary, covering some emissions but not others (it is impractical, for example, to extend it to the personal and household sector, including motoring). For those industries where it does apply, it is anti-competitive, since permits are issued to existing emitters, and not to new entrants, who have to purchase them from the market. In general, the administrative allocation system scores badly on transparency, and lends itself to lobbying, corruption and abuse of one kind or another. This is even more pronounced in an international scheme, when each government is under pressure to allocate generously to its own national emitters. Another problem is that it injects an artificial volatility into the price of energy, making rational investment decisions more difficult – not least any decision to invest in low-carbon or non-carbon energy. And, inevitably, the ethereal nature of the commodity being traded makes it particularly hard to police.

The only substantial emissions trading scheme so far attempted, the European Union's Emissions Trading Scheme (ETS), exhibits all these fundamental flaws – and indeed several more, as recent studies have shown.[19] In practice, it has done nothing to reduce emissions, and merely awarded subsidies to selected emitters. In theory, some of the disadvantages of the scheme could be avoided if the emissions permits were auctioned, rather than given away, but the design of an auction to cover all emitters – including the personal sector – and extending it internationally would be mind-bogglingly complex and contentious, if indeed it could be done at all. Which is no doubt why, for the new and allegedly improved second (2008–2012) phase of the ETS (the first phase is universally agreed, except by those who have made money out of it, to have been a farce) the EU has decided that 98.5% of the permits should be allocated and only 1.5% auctioned.[20]

The Clean Development Mechanism (CDM) set up under the Kyoto agreement, and with which Europe's ETS is linked, is no better. The idea is that, if a developed country with a Kyoto target finds it too difficult, or too costly to reduce its emissions, it can buy 'certified emissions reductions' (CERs) from developing countries instead. The certification is theoretically carried out by the United Nations, which has to satisfy itself that the reduction is genuinely additional (that is, that it would not have occurred anyway) and that (if, for example, it is achieved by shutting down a particular power plant) it is not offset by an increase in emissions elsewhere. In practice, the system is impossible to police, and newspaper investigations have revealed the CDM to be little more than a massive scam.[21]

It is also highly lucrative, both for the entrepreneurs who have discovered a new business opportunity, and for the companies and governments in the developing world who have organized themselves to take advantage of it. The CDM market is currently dominated by China. Under the Montreal Protocol the production of chlorofluorocarbons (CFCs), used principally in refrigeration, is meant to be phased out to protect the ozone layer, and in Europe and the United States it has been, but China has been much slower to do this. And CFCs are also powerful greenhouse gases – tens to thousands of times as powerful as carbon dioxide (depending on the precise nature of the gas). As a result, Chinese CFC producers are able to earn huge sums of money by selling 'certified emissions reductions' gained from reducing their CFC output below what it might otherwise (notionally) have been. So substantial is this revenue flow that the Chinese government has imposed a special tax on it, the proceeds of which can then be used to help finance its massive coal-fired power station programme.

According to one prominent environmental campaigner, 'Entrepreneurs in India and China have made billions by building factories whose primary purpose is to produce greenhouse gases, so that carbon traders in the rich world will pay to clean them up.'[22] Whether this is so or not, the incentive is clearly there: there are a growing number of industrial firms in the developing world for whom CDM revenues are their main source of profit.

For the sake of completeness, it should be noted that, in addition to the CDM, the Kyoto accord set up another mechanism, Joint Implementation (JI), under which countries with Kyoto targets which they more than fulfil can sell the resulting 'carbon credits' to developed countries who have difficulty in achieving their Kyoto targets, thus relieving them of the need to do so by reducing their own emissions. As has already been noted, the only Kyoto signatory whose emissions are running well below its target is Russia, thanks to the collapse of uneconomic Soviet-era heavy industry during the 1990s (the Kyoto targets, it will be recalled, were in terms of a 1990 benchmark).

So far, little has happened under JI, as Russia has been very slow to put in place the domestic regulatory structure required under Kyoto, but it expects to have this done during 2008, and is set to earn tens of billions of dollars by selling its carbon credits – affectionately known in the trade as 'hot air' – which it will then doubtless invest in the modernization and development of its important oil and gas industry. Indeed, it is more than likely that it was the prospect of this bonanza that finally decided the Russian leadership, after much hesitation, to ratify Kyoto in 2006.

In short, it is not going too far to conclude, looking at all these schemes, that the only practical effect of the Kyoto process has been to create what is fast becoming one of the biggest scams on the planet. Not that this has prevented Mr Yvo de Boer, the head of the UN Framework Convention on Climate Change, the IPCC's umbrella body, from suggesting that perhaps the developed world should forget all about cutting back their own emissions, and instead rely entirely on the (UN administered) CDM.[23]

There is also another, related, scam: carbon offsets. This is a wholly private sector response to the climate (in the non-literal sense) which the politicians and scientists have created. It has both a corporate and a personal dimension; corporations can promote themselves as 'carbon-neutral' by claiming to have purchased 'offsets' in the form of either emissions reductions elsewhere, or of CO_2 absorption through the planting of trees. And individuals can salve their consciences by doing the same, for example by offsetting their so-called 'carbon footprint' whenever they fly. In the case of individuals, the schemes on offer characteristically involve tree-planting.

Not surprisingly, newspaper investigations have revealed this to be largely, if not entirely, a scam.[24] The trees that have allegedly been planted may not have been; if they have been, they may well have been planted in any event, and either way their carbon absorption is notional, unverified, and at best, some way into the future. But in many ways carbon offsets can be regarded as a victimless scam. In the corporate sector, companies feel that the PR benefits of being able to parade themselves as being carbon-neutral are well worth the price they pay, and to investigate the *bona fides* of those from whom they purchase the offsets is far more trouble than it is worth.

For individuals, the equation is even clearer: those who feel guilty about travelling by air can, at relatively modest cost, assuage that guilt; the airlines that promote the offsets can ensure that otherwise conscience-stricken customers continue to fly; and a growing number of enterprising businessmen can make an excellent living out of the (at present perfectly legal) scam. Moreover, it does infinitely less harm to the economy than a genuine reduction in emissions is likely to cause.

In many ways, it resembles nothing so much as the sale of indulgences by the mediaeval church. This is nowadays regarded as a reprehensible practice; but perhaps, bearing in mind its 21st century equivalent, that is too harsh a verdict.

But for those who seriously wish to put in place a least-cost means of genuinely reducing carbon dioxide emissions, the remedy is clear: it is to forget about the ETS, CDM, CERs, JI and all that, and simply impose a carbon tax across the board. This, unlike cap-and-trade, genuinely is a market solution. We already have a fully-functioning energy market, which taxation would make use of. The superimposition of a market in emissions permits does nothing to improve the resource-allocation merit of the market system: at best it merely makes it less efficient, by adding a costly new infrastructure. Moreover, introducing a carbon tax, and standing ready to increase it, is the only practical means of discovering how expensive carbon needs to be in order to stimulate the changed behaviour necessary to stabilize emissions, on the demand side and the supply side alike, if that is the objective. It is only on the basis of this sort of information that rational policy decisions on emissions cutbacks could begin to be taken, should this be considered necessary.

It is true that there would be problems about securing international agreement on a tax increase. But the European Union has already acquired a competence in the indirect tax field – Value Added Tax, for example, is governed largely by various EU Directives; and I well recall that, when Chancellor, I was ordered by the European Court to change the UK's system of excise duties, which allegedly protected beer at the expense of wine (I conformed by increasing the beer duty very slightly and reducing the wine duty by considerably more). And, as we have already seen, outside the European Union, an international emissions agreement is in any event not on the cards.

No: the real reason why governments shy away from the tax route is that, while transparency is desirable in economic terms, for political reasons it is the last thing they wish to see. Obfuscation is far more attractive, not to mention the creation of a business community with a vested interest in the trading system, even if it is largely a scam, which can be counted on to support government policy on that account. Moreover 'tax' is an honorary four-letter word. Even though the entire yield of a carbon tax could (and indeed should) be used to reduce other taxes, governments fear that a carbon tax, with its effects on air fares and motoring costs, might prove unpopular. And the last thing they wish to have to reveal is the true cost of cutting back on CO_2 emissions.

Nonetheless, if we are serious about all this, estimating the true global cost of drastically cutting back on CO_2 emissions is precisely what we have to try and do. And the IPCC, in its 2007 Report, does indeed have a go at it. In what it describes as the 'Second Order Draft' of the relevant 'Summary for Policy Makers' – that is to say, the final draft before the report itself – it suggests that to stabilize CO_2-equivalent concentrations in the atmosphere at 550 parts per million by volume (they are at present around 430 ppmv) by 2050, would cause a loss of between 1% and 5% of that year's global GDP.[25]

This was changed in the final report to an estimate that to stabilize CO_2-equivalent concentrations in the atmosphere at between 535 ppmv and 590 ppmv by 2050 would cost anything from slightly less than nothing, to 4% of that year's global GDP.[26] It is not clear why this last-minute downward revision was made, but the suspicion is bound to be that the IPCC's directorate may have been concerned that to incur a cost of 1–5% of global GDP in 2050 in the hope of preventing at most a cost of 1–5% of GDP in 2100 (this, it will be recalled, was the IPCC's estimate of the likely cost of 4°C/7.2°F of further warming by then) may not have appeared a particularly compelling bargain.

In any event, even the mid-point of these inevitably conjectural ranges, some 2–3% of global GDP, is a very substantial sum indeed. Moreover, the final IPCC Report explicitly concedes that 'the costs and benefits of mitigation … are broadly comparable in magnitude'.[27] This is in stark contrast to the politically-inspired Stern Review, as indeed is the IPCC's careful agnosticism over whether 'later and less stringent mitigation' or 'earlier and more stringent mitigation' is more 'economically justified'.[28]

Again, even if it is indeed the case that the costs and benefits of mitigation are broadly comparable in magnitude, it is clear, as we have seen, that in practice the costs would be borne, in the first instance, entirely by the people of the developed world – including inevitably the poor people of the developed world – whereas the benefits, such as they are, would accrue in due course to the people of the world as a whole. Even those most concerned to help the developing world will wish to consider whether this massive and indiscriminate transfer, which would dwarf existing global aid flows, is the best way of doing so. This is, in effect, the point that Bjorn Lomborg has been eloquently making for many years now.[29]

As Chancellor of the Exchequer some twenty years ago, I launched the first concerted poor-country debt forgiveness initiative, subsequently known as the Toronto terms (it was agreed at the 1988 G7 Summit at Toronto).[30] If anyone had then argued that the best way to help the poorest of the poor was to make the world a colder place, I would probably have politely suggested that they see their doctor. It makes no more sense today than it would have done then.

But the fundamental question, when comparing the costs and the benefits of mitigation – even if we accept the conventional wisdom so far as the science is concerned, and even if we assume that a global agreement is attainable, however unlikely that may seem – is that posed at the end of Chapter 2. That is, how great a sacrifice is it either reasonable or realistic to ask the present generation, particularly the present generation in the developing world, to make, in the hope of avoiding the prospect that the people of the developing world, in a hundred years time, may not be 9.5 times as well off as they are today, but 'only' 8.5 times as well off?

Intuitively, the answer is clear: not a lot. And this is assuming that the huge cost of mitigation measures would indeed bring the benefits the IPCC projects, which may well not be the case. But since the prevailing orthodoxy is that tough measures are urgently needed, a more thorough analysis is required. Moreover this analysis needs to take fully into account our attitude to risk, to uncertainty, and indeed to ethical considerations. It is to these issues we must now turn, before finally reaching a conclusion.

Chapter 7

Discounting the Future: Ethics, Risk and Uncertainty

There is a standard way of measuring future benefits against present costs; that is, to apply a rate of discount to the future. For jam tomorrow is not worth as much as jam today, nor is the possibility or even the probability of jam tomorrow worth as much as the certainty of jam tomorrow. Moreover a pot of jam is worth more to the poor man than to the rich man. So, for long-term projects, the richer future generations are expected to be, the higher the appropriate discount rate.

Applying a rate of discount to the future is what businessmen do when deciding whether or not to make a business investment, and it is what governments do when assessing public sector investment projects. We invest money now, whether in physical or human capital, only because we expect to earn a return over time. The average return, over and above inflation, for the economy as a whole (including non-market returns) gives us the rate of discount that governments need to use in deciding what expenditures to undertake. At the very least, it should ensure consistency of treatment. The case of global warming may be particularly difficult, since the alleged benefits of mitigation now are much further off in time than is the case for the great majority of (but not all) conventional capital investment projects. But the same principles apply.

The IPCC itself concedes that 'Comparing mitigation costs with the benefits of avoiding climate change damages and other [alleged] co-benefits is very complex. This is caused by [among other factors] the sensitivity of benefit calculations for the assumptions made, such as about the discount rate.'[1] Clearly, the lower the discount rate, the more the future looms large. So the choice of discount rate is in fact critical in assessing which policies might make sense, and which clearly do not.

The IPCC is judiciously silent about what discount rate to adopt, but it would clearly need to be a very low one indeed to justify, even on its own assumptions about likely costs and benefits, a policy of taking radical action now to stabilize carbon dioxide concentrations in the atmosphere. The Stern Review, which confidently asserts that the costs of mitigation are likely to be roughly half the mid-point of the IPCC's (downwardly-adjusted) range – 1% of global GDP rather than 2% – and inflates the damages of global warming to a figure substantially greater than the IPPC's range of 1–5% of global GDP for a rise in temperature of 4°C/7.2°F, appears to assume a discount rate of at most some 2% in order to make the case for urgent and radical action.[2]

It is fair to say that the Stern Review's low discount rate has been roundly criticized by almost all the most eminent academic economists, from Sir Partha Dasgupta at Cambridge, to William Nordhaus, the doyen of global warming economics, at Yale.[3] Professor Dasgupta has pointed out that the Stern Review's assumptions imply that the present generation ought to be saving 97.5% of its income for the benefit of future generations, which he rightly describes as 'patently absurd', while Professor Nordhaus has also categorized the policy prescription of the Stern Review as 'completely absurd'. Professor Martin Weitzman of Harvard, too, has observed that 'the *Review*'s radical policy recommendations depend upon controversial extreme assumptions and unconventional discount rates that most mainstream economists would consider much too low'.[4] And it has been demonstrated that with a higher, more normal discount rate, the argument for radical action over global warming now, on conventional cost-benefit calculations, collapses completely.[5]

An immediate problem is that of comparing the cost of mitigating global warming with the cost of other, competing, projects. In the UK, the Treasury requires a public sector project to make sense on the basis of a 3.5% discount rate if it is to be allowed to go ahead (in my day the hurdle rate was 6%: the justification for the subsequent reduction is highly dubious). The private sector, of course, requires a considerably higher rate than this. Moreover, what is at issue here is not a UK, but a global project. And in the developing world, both the greater growth potential and the greater risk mean that the discount rate hurdle is appreciably higher. In India, for example, as the economist Deepak Lal has pointed out in a telling and succinct critique of Stern,[6] a normal rate would be 7%, while the World Bank customarily uses a rate of some 8–10% in evaluating projects, even long-lived projects that are expected to last a hundred years and more, such as major infrastructure or educational projects.

Is there, then, something so special about global warming that it should receive this exceptionally indulgent treatment? If so, and it is not at all clear why it should be so, unless perhaps we adopt the extreme alarmist position, which neither the IPCC nor the evidence supports (and it could be argued that, if we do take the extreme alarmist position, then it is already too late to prevent disaster[7]), where do we draw the line? Should the assessment of nuclear power projects, which provide largely carbon-free electricity, be based on a 2% discount rate? And what about adaptation projects, such as flood defences; should these also be evaluated on a 2% discount rate? If not, why not? And if adaptation projects are to be given this privileged treatment, on what grounds do we distinguish, for example, between measures to protect people from diseases that might be exacerbated by global warming, and measures to protect them from diseases unrelated to global warming?

Moreover, wherever the line is drawn, the practice of using different hurdle rates for different projects destroys the whole purpose of having a hurdle rate in the first place. It is bound to lead to major public spending decisions being made – even more than they otherwise would be – on the basis of short-term political considerations, with a consequent misuse of resources and of the taxpayers' money.

It is clearly a complete muddle, which, needless to say, the Stern Review does not even attempt to address. Instead, it seeks refuge in what it believes to be the moral high ground, insisting that, at the end of the day, the issue is all about ethics, and declaring (in bold type) that 'if you care little about future generations, you will care little about climate change. As we have argued, that is not a position which has much foundation in ethics and which many would find unacceptable.'[8]

This does not, of course, dispose of the problems to which I have just referred; there is an ethical dimension to most, if not all, decisions, not just those related to global warming. But the ethical dimension of global warming is certainly important. The Stern Review's treatment of this, however, is profoundly unconvincing. What it does, in effect, is to insist that the only possible ethical position is to consider the welfare of all future generations, stretching ever further out into the mists of time, as being just as important as the welfare of the present generation. It is this assumption which largely (although not entirely) generates its very low discount rate.

This is certainly one possible ethical position, and Stern is not the first to hold it (the utilitarians appear to have done so, too). But, equally, it is certainly not the only one, nor is it the way most of us behave in real life. We care about the welfare of our children and our grandchildren, but we do not normally lose sleep over the welfare of our grandchildren's putative grandchildren, nor make financial provision for them. Nor have we any reason to feel guilty about that. As David Hume pointed out, getting on for three hundred years ago in his classic work *A Treatise of Human Nature*, morality is based on human behaviour and human affection:

> A man naturally loves his children better than his nephews, his nephews better than his cousins, his cousins better than strangers, where everything else is equal. Hence arise our common measures of duty, in preferring the one to the other. Our sense of duty always follows the common and natural course of our passions.[9]

Or, as Ian Little, Britain's most distinguished living authority on welfare economics has put it, explicitly following Hume:

> A moral code evolves from the customs and conventions needed to permit human beings to live together in a peaceful and productive community. Within such a community a person values the welfare of friends, and usually relatives, higher than that of strangers; and probably that of co-citizens higher than that of foreigners. One may think of 'social distance' as determining differences of moral concern. Time is a determinant of social distance.

Little concludes that 'discounting future welfare does not deserve the antipathy that it has received in the literature'.[10]

To take the extreme case, while it may at first blush seem heartless to say that the welfare of those living in the next millennium is of no consequence, to take decisions on the basis that it is every bit as important as the welfare of the population of the world today would be palpably absurd. In particular, it would mean that, since there are so many future generations involved, we ought to make a big sacrifice now in order to confer on each distant (and incidentally better off) generation a trivial benefit, if we believe we know how to do so. That is the ethics of the Stern Review.

Indeed, to care no more about the problems, many of them acute, of those alive today than we do about the possible problems of remote future generations would be judged by many (in my opinion rightly) to be profoundly *im*moral – or, in the words of the Stern Review, 'unacceptable'. To put it another way, it is not that we do not care about distant generations, it is that we *do* care about the present generation, and indeed about our children's generation, who would also pay a high price if the world were to slash carbon dioxide emissions to the extent we are told is required.

The reader may recall the absurd Mrs Jellyby in Charles Dickens' *Bleak House*,[11] the 'Telescopic Philanthropist' who was so concerned with good works in Africa, and with the Brotherhood of Humanity in general, that she neglected her own children. The self-proclaimed ethical basis of the Stern Review's discount rate is little more than intertemporal Jellybyism.

But while the alleged ethical basis of the Stern Review's ultra-low discount rate, on which its argument for urgent radical action entirely rests, is unconvincing, is it possible that, as Weitzman has argued, the review may still be right, albeit for the wrong reason?[12] In a nutshell, Weitzman is concerned, as are some others[13], that it cannot be ruled out, however unlikely it may be, that man-made global warming may have consequences that are so catastrophic that it is worth paying almost any price to try and avert that risk.

At the present time I do not find this argument compelling. As we saw in Chapter 4, there is no hard scientific evidence to suggest that catastrophes of this kind may be in the offing, nor is it plausible that they could suddenly emerge without warning. Moreover there is almost certainly a viable measure to keep in our back pocket for use should the highly unlikely threat of catastrophe ever loom large (as we shall see in the final chapter).

We do indeed have to decide how risk-averse we are. This, incidentally, is another reason why a Kyoto-style global agreement is a fool's errand. Not only is there the impossibility of agreeing how to share the burden between the major nations of the world, but there is also the fact that risk aversion is not uniform throughout the world. Different cultures embody very different degrees of risk aversion. In general, and other things being equal, the richer countries of the world will tend to be more risk-averse than the poorer countries, if only because we have more to lose.

It is not for us to tell others how risk-averse they ought to be, or how they should interpret that notably elusive and highly subjective concept, the so-called precautionary principle. Of course, even within a country, there will be different degrees of risk aversion. But where collective action is needed (over national defence, for example) it is the government's job to strike a balance which is generally acceptable. But world government – happily, because the disadvantages would greatly outweigh the advantages – is a horse that is never even going to come under starter's orders.

Moreover, to the extent that we are risk-averse, we have to be risk-averse consistently. That is to say, it makes no sense to be more risk-averse to the possible adverse effects of global warming than we are to the other risks that face us (including the 'risk' that global warming, if indeed it continues to happen, has little or nothing to do with carbon dioxide emissions, and that mitigation policies turn out to be the greatest misuse of resources the world has ever known). Perhaps the most important application of the so-called precautionary principle is to the precautionary principle itself, otherwise we may find ourselves doing very foolish things in its name.

Without risk-taking there is no human progress, whether it is the risk taken by entrepreneurial businessmen, or the risk taken in scientific development (the GM foods saga is a particular case in point). Certainly risk always needs to be assessed and monitored, but to take policy decisions on the basis, in an inherently uncertain world, of the worst possible case, is not rational precaution, but irrational alarmism. It is probabilities, not possibilities, that should be our guide.[14]

At first glance the suggestion that perhaps we ought to be prepared to sacrifice up to, say, 5% of today's GDP as a kind of insurance policy[15] against the remote risk of a catastrophe in the distant future, may have some plausibility (even though, given the prospective rise in living standards over time, and insofar as the insurance analogy is valid, it is rather like suggesting to someone today that he should take out an insurance policy now, and start paying the premiums, against the risk of his great grandchild's yacht capsizing).[16]

But the greatest flaw in this approach is that there are a large number of unlikely but possible catastrophes waiting to happen at some time in the future, of which runaway global warming is only one; and there is clearly no rational case for focusing on that one alone. To do so would be a particularly purblind example of tunnel vision. Others include a nuclear holocaust, the mother and father of all flu pandemics (whether avian or any other kind), a devastating asteroid collision with the earth, and indeed the onset of a new ice age. If, in the name of risk aversion and the precautionary principle, we were to sacrifice, say, 5% of today's GDP to try and guard against each of these, this would mean sacrificing up to 25% of today's GDP in the hope of avoiding all of them (and there are doubtless others, too), even though their combined likelihood is still very small. This would be plainly absurd.

In his book *Our Final Century*[17], Professor (Lord) Martin Rees, the current President of the Royal Society, maintains that mankind has at best a 50% chance of surviving the present century. I have no idea whether or not he is right, but one thing is clear: although, as an all-purpose Cassandra, he does not fail to mention global warming, the more serious and more imminent threats he sees are from biotechnology, in the sense of what he terms bioterror and bioerror; from nanotechnology running amok; and above all from nuclear conflict, arising in particular from the proliferation of nuclear weapons, coupled with the fact that technological development has brought the means of devastation within the reach of even relatively modest terrorist groups. This last clearly threatens us well within the centuries-long time scale that preoccupies the Stern Review and the global warming scaremongers in general.

Nuclear proliferation in an age of terrorism is indeed a real threat to the planet, which probably needs to have first claim on what might be termed our risk aversion resources. In a world of inevitably finite resources, we cannot possibly spend large sums on guarding against any and every possible eventuality in the future. Reason suggests that we concentrate on present ills, such as poverty and disease, and on future dangers, such as nuclear conflict and terrorism, where the probability appears significant – usually because the signs of their emergence are already incontrovertible. The fact that a theoretical future danger might be devastating is not enough to justify substantial expenditure of resources here and now, particularly since there are many other such dangers wholly unconnected with global warming.

That, in the real world, resources are finite, and we have to prioritize, is unpalatable to some, but the constraint is inescapable. This is particularly so in a genuine democracy, where a government also has to balance and respect the differing priorities of the people it serves. It is striking, for example, that the United States, for all its great wealth, and for all the great importance that the Bush administration attached to the Iraq adventure, has found itself severely resource-constrained in conducting it. This would have been the case even if the whole adventure had enjoyed universal popular support.

Moreover, the politicians need to be honest with the people, and tell them the truth. If they believe that we need to cut back drastically on carbon dioxide emissions today, at considerable cost and disruption to our way of life, not because there is any real likelihood of significant harm from global warming, but because there is a remote risk of major disaster at some time in the distant future, they should make the case explicitly in those terms, and in no other.

Chapter 8

Summary and Conclusion: A Convenient Religion

It is time to reach a conclusion.

As we have seen, perhaps the most surprising fact about global warming, given that nowadays pretty well every adverse development in the natural world is automatically attributed to it, is that – despite carbon dioxide emissions rising faster than ever – it is not, at the present time, happening.

There was, indeed, a rise in the world's temperature of about half a degree centigrade during the last quarter of the 20th century, and its effects are apparent, but even the Hadley Centre for Climate Prediction and Research, Britain's world-renowned citadel of the conventional global warming wisdom, has now conceded (and in the light of the recorded temperature series for the first seven years of this century it could hardly not do so) that since then there has been a standstill. It officially expects global warming to resume in 2009 or thereabouts. Maybe it will; we shall see.

But the fact that this lull was not predicted by any of the immensely complex computer models which embody the conventional wisdom, is clear evidence (and as we have seen, by no means the only evidence) that the science of what determines the world's temperature is far from 'settled'. The earth's climate is determined by hugely complex systems, many important aspects of which are not at all well understood. Reliable prediction is impossible.

Fortunately, despite the seemingly endless media hype,[1] opinion surveys suggest that a clear majority of ordinary people, even in the UK where politicians of all parties all sing from the same politically correct hymn sheet, instinctively sense that this is so.

Needless to say, the fact that the science is not settled does not mean that we know nothing. We do know that, through the so-called greenhouse effect, carbon dioxide in the atmosphere warms the planet, and that, ever since the industrial revolution, man has been adding to the amount of carbon dioxide in the atmosphere – and continues to do so – by his reliance on carbon-based energy. So it might be reasonable to suppose that, other things being equal, the world is likely to get warmer.

But that is where the uncertainty takes over. In the first place, other things, including the natural forces that influence the earth's temperature, are most unlikely to be equal. And in the second place, even if they were equal, there is considerable dispute among climate scientists about how great the consequent warming effect of increased CO_2 concentrations in the atmosphere would be.

Again, as we have seen, the conventional wisdom, as spelled out in the 2007 Report of the Intergovernmental Panel on Climate Change (but before the early-21st century warming lull was officially acknowledged), is that between now and the year 2100, we can expect a warming of between 1.8ºC/3.2ºF and 4ºC/7.2ºF.

The notion that, even if it were to occur, this would be such a disaster that we need to take radical action now to cut back on carbon dioxide emissions in order to 'save the planet' has been seen to be wholly without foundation. Not surprisingly, gradual and moderate warming brings benefits as well as incurring costs. These benefits and costs will not, of course, be felt uniformly throughout the world; the colder regions of the world will be more affected by the benefits, and the hotter regions by the costs. But overall, it is far from clear that the inhabitants of the planet as a whole would suffer a significant net cost, or indeed any net cost at all.

The IPCC's own rather more pessimistic assessment is seen to rest partly on a reluctance to acknowledge sufficiently the benefits of global warming (although it does explicitly reckon that up to 3ºC/5.4ºF of warming would be helpful to world food production), but more importantly on its wholly inadequate appreciation of mankind's ability to adapt to gradual change, and thus to minimize the costs, as has happened throughout the ages. This is, of course, something that, in a competitive market economy, will to a considerable extent happen naturally, spontaneously and autonomously, without any need for government intervention. But where 'public goods' are concerned, government intervention will indeed be required, should the need arise.

One of the central messages of this book is that, in the light of the uncertainty that exists about the science, and the inevitable uncertainty there is about the future in general, it must make more sense to rely on autonomous adaptation, buttressed where necessary with positive policy measures to assist it, than to pay a very heavy price to try and secure a drastic reduction in emissions without even any realistic likelihood of achieving this.

But even on the basis of the IPCC's flawed economic assumptions we find that the existential threat to the planet, the disaster which we must do all in our power to avert, is merely that living standards in the developing world in a hundred years time are projected be 'only' some 8.5 times as high as they are today, instead of some 9.5 times as high, without the alleged ravages of global warming. 'Save the planet' must surely be a strong contender for the most ludicrous slogan ever coined.

Nor does either scientific theory, or the hard evidence on the ground, or the considered views of reputable climate scientists, support the Gore-style views of alarmist politicians that we are on the road to catastrophe, as a result of the planet reaching some irreversible 'tipping point'.

Moreover, even if, to err on the safe side, we were to accept both the conventional wisdom that anthropogenic global warming presents a serious problem, and also the IPCC's flawed economic analysis, a policy of cutting back sharply carbon dioxide emissions in an attempt to stabilize CO_2 concentrations in the atmosphere, is not a sensible course to take. For what would be required (according to the conventional wisdom) is a reduction in worldwide emissions, which at present continue to rise steadily, of at least 70%.

In this context, it is clear that the feelgood measures so popular among parts of the Western middle classes are trivial to the point of irrelevance. A major change in our way of life would be required, brought about by a very substantial rise in the price of energy, both so that we use much less of it and so that non-carbon energy becomes economic. Quite apart from the public resistance a policy of this kind would engender, the economic cost of implementing it would be likely to far exceed the benefits it is hoped to secure.

But in any event, even if we in the West wished to follow this route (and in fact, outside the European Union, the official support for it is largely lip-service), it would be possible – and this, too, is not in dispute – only on the basis of a binding Kyoto-style global agreement on emissions limits. And as we have seen, this is not on offer.

The principal, although not the only, reason why a binding global agreement to cut back drastically on emissions is not on offer is that the major developing nations, notably China and India, have made it clear that, at least for the foreseeable future, they will have no part of it. They are fully justified in adopting the position they do. In the first place, their emissions per head of population, although now rising rapidly, remain well below those in the West and will continue to do so for several decades to come. In the second place, and more importantly, their overriding concern is to improve the condition of their people, still mired in mass poverty, by the fastest rate of economic growth they can muster, which a sharp rise in the price of energy would seriously impede.

At the heart of this issue is the question of how great a

sacrifice it is right to ask the people of this generation and the next to make in the hope of benefiting future generations, a hundred, or two hundred years, or indeed a thousand years hence, who in any event are likely to be many times better off. The Stern Review is right in maintaining that this is at root an ethical issue, but the ethical issue is not just about how much we care about distant future generations; it is also about how much we care about the present generation, not least in the developing world, and its children. Certainly, for the governments of those countries, the question of how great a sacrifice the present generation and their children should make, in terms of unnecessary poverty, malnutrition, disease and premature death, in the hope of benefiting substantially better off generations a hundred or two hundred years hence, is not a difficult one, either in ethical or indeed in political terms.

Theoretically, to be sure, the governments of the developed world could agree among themselves to increase the taxes on their own people by whatever it takes to enable them to bribe China, India, Brazil and the rest of the developing world sufficiently to make it worth their while to accept a sharply increased price of carbon. But anyone who believes this to be a politically realistic way forward need not bother about saving the planet: they are already living on a different one.

So does all this mean that we should do nothing about global warming? Not quite, although doing nothing is better than doing something stupid. But there are, in fact, a number of sensible things that could be done – most of which, happily, are to a considerable extent already being done.

Plainly, we need to monitor as accurately as we can, and without preconceptions, what is happening to temperatures around the globe, and what is happening to the natural phenomena which changes in temperature might affect. This is all the more important given the significant variation that has already been observed in the behaviour of ice sheets and other natural phenomena. We need to try and understand much better than we do, the various factors, from clouds to cosmic rays, that either undoubtedly do, or might, influence the climate. This requires the funding of serious research of all kinds. It is a scandal that, at the present time, reputable climatologists who are sceptical of the current anthropogenic global warming orthodoxy find it very difficult to attract funding, or to get their papers published, and indeed are all too often vilified.[2]

It is clearly right that there is already substantial and growing research and development expenditure, particularly in the United States, both by the private sector and by governments, on a wide range of reduced-carbon and non-carbon sources of energy, in the hope of bringing forward the day when at least some of these technologies may become economic.[3] At least this goes with the grain. Whereas a nation which (like the UK if the government means what it says), cuts back on its emissions in the near future, is bound to lose out competitively, a nation which achieves a technological breakthrough is likely to benefit competitively – even if, as ought to be the case, there is rapid technology transfer. Equally important, in this context, is the need to overcome the irrational hostility to nuclear power.

At the same time, subsidies for carbon-based energy, particularly in the developed world (including not least the European Union), should be eliminated without delay.

Again, it clearly makes sense to press ahead with research and development in technologies that might assist the process of adaptation should that be required, as well as having practical utility even in the absence of warming. Desalination is one such technology, and there are several others, as will have been apparent from the discussion of adaptation in Chapter 3.

Another form of R & D which is rightly taking place at the present time, although so far only in the United States, involves what has become known as geoengineering; that is, the technology of cooling the planet, in relatively short order, should the need ever become pressing. In a sense this is the last piece of the jigsaw; the rational response should the highly unlikely threaten to occur – that is, a degree of warming with consequences beyond man's capacity to adapt.

The front-runner here is the idea of blasting aerosols into the stratosphere, so as to impede the sun's rays. Such grand schemes obviously need to be approached with caution, but it is striking that they have gained the support of scientists of the eminence of the Nobel Prize-winner, Professor Paul Crutzen.[4] Another scientist who has done, and is continuing to do, important work on this, is Professor Ken Caldeira of the Carnegie Institution's Department of Global Ecology at Stanford University, California.[5]

This proposal is not nearly as far-fetched as it seems. In essence it reproduces what happens naturally when large volcanoes erupt. The most recent such eruption was that of Mount Pinatubo in the Philippines, in 1991, which ejected some 10 million tonnes of sulphur into the stratosphere, and is generally agreed to have led to a distinct cooling of the earth in 1992 and 1993 of at least 0.6°C/1.1°F, with no recorded adverse side effects of any kind (sulphur at lower levels of the atmosphere is indeed harmful to man, but not so high up in the stratosphere). The adverse side effect feared by some, should this form of geoengineering ever be put into effect, is serious damage to the planet's ozone layer. There is, however, no evidence that sulphate particles ejected into the stratosphere by volcanic eruptions have had this effect, and Crutzen discounts such fears. Since it was for his work on the ozone layer that he was awarded his Nobel Prize, this view carries weight.[6]

It is also worth noting that this process would cool the planet without losing the beneficent fertilization effect of the increase in atmospheric carbon dioxide – the best possible outcome for plant growth, as it happens. And although, should geo-engineering be called for (as, to repeat, seems highly unlikely), it would be desirable to secure global agreement to engage in it, it would of course work equally well in the absence of any such global accord.

I conclude that it is clearly a worthwhile precautionary policy to spend government money, which the United States appears prepared to do, on further research into geoengineering of various kinds, and to develop the capability (where this does not already exist) to put it into practice in the unlikely event that the need arises.[7]

More importantly, there is of course the need to do whatever is required to adapt to a warmer planet, should the late 20th century warming, which has for the time being paused, soon resume, as the majority of climate scientists are currently predicting. For the most part this can and will happen spontaneously and autonomously, just as mankind has always adapted to the environment around him, wherever he lives, without any need for government intervention. But there are some areas – what the economists call the supply of 'public goods' – where governments do need to stand ready to act. The provision of adequate sea and flood defences is the most obvious example.

Moreover, as we have seen, even though the IPCC's projected warming over the next hundred years, if it occurs, may well not be harmful overall, there would be losers in the warmer regions of the developing world. Should this seem likely to occur, I believe we have a clear moral obligation to help them. It is true that the record of overseas aid in promoting economic development is very disappointing. This is because economic progress depends on a free, open and well-functioning market economy, which in turn depends on an institutional and cultural infrastructure – in particular the rule of law, including the enforcement of contracts, respect for property rights and, preferably, a restrained level of corruption – which no amount of aid can bring about, and which only good governance can. As Adam Smith succinctly put it:

> Little else is requisite to carry a state to the highest degree of opulence from the lowest barbarism, but peace, easy

taxes, and a tolerable administration of justice; all the rest being brought about by the natural course of things.[8]

But that is no argument against assistance in, for example, the building of effective sea defences. Of course it would cost money. But quite apart from our moral obligation, it is only a minuscule fraction of what it would cost to attempt, by substantially cutting back on carbon dioxide emissions, to control the global temperature.

Finally, I believe that a case can be made for the introduction of an across-the-board carbon tax, initially at a relatively low level, *provided (and this is of the first importance) that the proceeds of the tax are fully returned to the pockets of the people, pound for pound or dollar for dollar, by a reduction in other taxes, such as income tax*.

If the objective is seriously to reduce global emissions, it would need to be a consumer-based tax, since in the globalized world economy industry is highly mobile, whereas individuals are far less so (although if it should lead to any 'leakage' to the developing world, this could always be seen as a more useful form of overseas aid than most). An industry-based tax, of course, would make it easier, and less costly, for the UK to preen itself on reducing its carbon dioxide emissions, as energy-intensive industries migrated overseas. The observations of China's Mr Qin Gang, noted in Chapter 5, make this point. Either way, it would give the UK an opportunity to propose, for a change, a constructive European initiative to its European Union partners.

The case for a carbon tax is essentially twofold. In the first place, as Colbert, the great 17th century reformer of the French tax system is reputed to have said, the art of taxation is to pluck the goose so as to obtain the largest amount of feathers, with the smallest possible amount of hissing. It is on this basis that, for many years, I and my predecessors and successors as Chancellor of the Exchequer in this country (and many of our counterparts elsewhere in Europe) have used high-sounding health arguments to justify raising substantial revenues from tobacco taxation, always taking care not to pitch the duty so high that too many people gave up smoking, causing the tax yield actually to diminish. In the same way, if people like to feel that they are helping to save the planet by paying a carbon tax, they should not be deprived of the opportunity to do so.

In the second place, levying a carbon tax, initially at a low level (it could always be subsequently increased in the light of experience, should we so wish), is the only practical way of getting an indication of what it might take to change behaviour sufficiently seriously to cut back on carbon-based energy consumption, if that is what we decide to do. However, what needs to be absolutely clear is that there is no point in increasing it beyond the level that taxpayers are content to bear, bearing in mind of course that they will be recompensed by equivalent reductions in other taxes. I suspect that this would prove to be a much more modest increase than the true believers would like to see. If so, it should be noted that the popular resistance would not be to the tax burden, which *ex hypothesi* would not have risen at all, but to the higher energy price and its consequences – which would apply however that price rise is generated.

There may well be no objective need for a carbon tax: the case for it is essentially pragmatic. Even if the damage done (if indeed there is any damage done) by increased atmospheric concentrations of carbon dioxide is no greater than the economic and indeed human cost of forced decarbonization, if the spirit of the age demands that something be done, if only as a gesture, to curb CO_2 emissions, then a carbon tax imposed on a strictly revenue-neutral basis would do no great harm. That is a good deal more than can be said for either a cap-and-trade system, which has been seen to be both undesirable in principle and a scam in practice, or the capriciously intrusive 'interventionist gimmickry' (to quote the *Financial Times*'

Martin Wolf) to which both Britain's political parties appear to be addicted.[9]

What is important is that the practical measures I have outlined in the last few pages represent the sum total of what we should be doing. It has to be said that this is not the easiest of messages to get across – not least because the issues surrounding global warming are so often discussed in terms of belief rather than reason.

In part, there may be a political explanation for this. With the collapse of Marxism, and to all intents and purposes of other forms of socialism too, those who dislike capitalism, not least on the global scale, and its foremost exemplar, the United States, with equal passion, have been obliged to find a new creed.

For many of them, green is the new red.

And those who wish to take power to order us how to run our lives, faced with the uncomfortable evidence that economic prosperity is more likely to be achieved by less government intervention rather than more, naturally welcome the emergence of a new licence to intrude, to interfere and to regulate: the great cause of saving the planet from the alleged horrors of global warming.

A major difference between the red and the green is that between optimism and pessimism. Marx, adopting and adapting Hegel's notion of historical inevitability, was fundamentally idealistic and optimistic. Society would go through various phases, culminating in the victory of the proletariat and the consequent withering away of the state, whose only historical function had been to oppress the proletariat. (Although the process was inevitable, it was nonetheless necessary to promote revolution to hurry it along. Marxism is nothing if not impatient.) The red left, however watered down its Marxist ideology may have become, retained his idealistic and optimistic view of man's social and material progress.

The green left, by contrast, is profoundly pessimistic; the

world is going to hell in a handcart as a result of the excesses of materialist capitalism. So far from believing in the future, it is attracted to a mythical pre-materialist and pre-capitalist past. What the red and the green do have in common, however, is a profound distaste for the liberal capitalist present, and an addiction to collectivist means of escaping from it. It is not hard to see how those who initially embraced the red left have shifted easily, when the practical embodiment of their idealism proved so disastrous, to the green left. Transparent realism is greatly to be preferred to both red optimism and green pessimism.

But there is something much more fundamental at work. I suspect that it is no accident that it is in Europe that eco-fundamentalism in general and global warming absolutism in particular, has found its most fertile soil; for it is Europe that has become the most secular society in the world, where the traditional religions have the weakest hold. Yet people still feel the need for the comfort and higher values that religion can provide, and it is the quasi-religion of green alarmism and what has been well described as global salvationism (of which the global warming issue is the most striking example), which has filled the vacuum, with reasoned questioning of its mantras regarded as little short of sacrilege.

Throughout the ages, something deep in man's psyche has made him receptive to apocalyptic warnings: 'the end of the world is nigh'. And almost all of us are imbued with a sense of guilt and a sense of sin. How much less uncomfortable it is, how much more convenient, to divert our attention away from our individual sins and reasons for guilt, arising from how we have treated our neighbours, and to sublimate it in collective guilt and collective sin.

Throughout the ages, too, the weather has been an important part of the religious narrative. In primitive societies it was customary for extreme weather events to be explained as punishment from the gods for the sins of the people; and there is no shortage of examples of this theme in the Bible, either – particularly, but not exclusively, in the Old Testament.

Nor have the old religions been slow to make common cause with the new religion of climate change. The Archbishop of Canterbury not so long ago told politicians that they would face 'a heavy responsibility before God' if they failed to act to curb global warming, and described the lifestyle of those who allegedly contribute most to global warming as 'profoundly immoral'. He added that 'if we look at the language of the Bible on this, we very often come across a situation where people are judged for not responding to warnings'.[10] (Whether it is theologically sound to equate warnings from the Almighty with those derived from computer models is not for me to judge.)[11]

Does all this matter? Up to a point, no. Fortunately, the gap between rhetoric and reality when it comes to global warming, between the apocalyptic nature of the alleged threat and the relative modesty of the measures so far implemented (not to mention the sublime disregard of international obligations solemnly undertaken), is far greater than I can recall with any other issue in a lifetime of either observing or practising politics. The explanation, of course, is that while fine words are cheap and probably politically attractive, the deeds to match them are anything but cheap and almost certainly politically unattractive. While the consequence in terms of political posturing may be distasteful, at least it has so far mitigated (to coin a phrase) the damage that would have been done had the more strident governments' deeds matched their extravagant words.

Moreover, unbelievers should not be dismissive of the comfort that religion can bring, even if some of us prefer to seek our spiritual solace in the music of Mozart, for example. If people feel better when they drive a hybrid car or ride a bicycle to work, and like to parade their virtue in this way, then so be it. (There is, however, something particularly unattractive about high-profile pop stars and the like telling the rest of us that we should not be flying to foreign destinations on holiday, whereas they need to do so for reasons of 'work'.)

And for political leaders in a democracy, the new religion is particularly convenient. It is not simply that they have discovered a wonderful diversion from their failings in more mundane matters, for which the voters may hold them responsible, although this is certainly a major factor. It also solves a deeper problem, which the economist and thinker Joseph Schumpeter to some extent foresaw more than sixty years ago, when he wrote that 'Capitalist rationality does not do away with sub- or super-rational impulses. It merely makes them get out of hand by removing the restraint of sacred or semi-sacred tradition.'[12]

The problem for political leaders in a free capitalist democracy is that, while capitalism is essentially opposed to authority, government requires authority. For a considerable period after Schumpeter made his observation, this was secured by a continuing deference to duly constituted authority, despite the decline of what he described as 'sacred or semi-sacred tradition'. But with the waning of deference of all kinds, the opportunity for political leaders to solve this problem by clothing themselves in the priestly garb of the new religion and proclaiming themselves the saviours of the planet, is too good to miss.

Nonetheless, the new religion of eco-fundamentalism and global warming presents dangers on at least three levels. The first is that it breeds an intolerance of dissent and reasoned argument that is both unattractive and dangerous. The attempt by the Royal Society, of all bodies, to prevent the funding of groups and organizations which openly doubt the alarmist creed of the new orthodoxy, on the grounds that they are 'providing inaccurate and misleading information to the public', is particularly shocking – and telling.[13] It is clearly undesirable that no young scientist, or young politician, dare question the new religion without severely damaging their career prospects, (indeed, I have been able to write this book only because my own career is behind me). Nor is it a coincidence that so many of the qualified scientists who publicly question the conventional wisdom are retired. The PC at the heart of the IPCC, as it were, is the most oppressive and intolerant form of political correctness in the western world today.

The second danger is that the governments of Europe may get so carried away by their own rhetoric as to impose measures which do serious harm to their economies. This is a particular danger at the present time in the UK.

And the third, and still greater danger, is that even if the voters prevent Europe's governments from going too far to damage their own economies, they may still cause great damage to the developing world by engaging in what might be termed green protectionism. The movement to make us feel guilty about buying overseas produce because of the 'food miles' involved is just one example of this. A more fundamental threat comes from the growing calls from luminaries such as the European Industry Commissioner, Mr Verheugen, for the imposition of trade sanctions against those nations which (quite rightly, as we have seen) will not agree to increase substantially the price of carbon within their borders – a threat which is even entertained by the lamentable Stern Review.[14] France's President Sarkozy has become an increasingly vocal advocate of this, for example declaring, on a visit to China a week ahead of the Bali gathering, that 'I will defend the principle of a carbon compensation mechanism at the EU's borders with regard to countries that do not put in place rules for reducing greenhouse gas emissions.'[15]

It should not need pointing out that a lurch into protectionism, and a rolling back of globalization, would do far more damage to the world economy, and in particular to living standards in developing countries, than could conceivably result from the projected continuation of global warming. But even if this danger can be averted, it is clear that the would-be saviours of the planet are, in practice, the enemies of poverty reduction in the developing world.

So the new religion of global warming, however convenient it may be to the politicians, is not as harmless as it may appear at first sight. Indeed, the more one examines it, the more it resembles a *Da Vinci Code* of environmentalism. It is a great story, and a phenomenal best-seller. It contains a grain of truth – and a mountain of nonsense. And that nonsense could be very damaging indeed. We appear to have entered a new age of unreason, which threatens to be as economically harmful as it is profoundly disquieting. It is from this, above all, that we really do need to save the planet.

Notes

Introduction

1. See P Ehrlich, *The Population Bomb*, 1968; also W & P Paddock, *Famine: 1975!*, 1967.

2. D Meadows et al., *The Limits to Growth,* 1972. The book sold more than 12 million copies worldwide.

3. See, for example, Peter Gwynne, 'The Cooling World', *Newsweek,* 28 April 1975, or at: http://brainterminal.com/common/images.php/newsweek-cooling-world.pdf

4. Robert Malthus, *An Essay on the Principle of Population,* 1798.

5. For a good recent account of the scare phenomenon in general, see *Scared to Death: The Anatomy of a Modern Madness,* Christopher Booker & Richard North, 2007.

Chapter 1: The Science – and the History

1. HL Paper 12-1 (session 2005-06). The present author declares an interest as a member of that Committee.

2. Counting heads in this area is in any case easier said than done. It is sometimes claimed, for example, that the scientific account published in the reports of the Intergovernmental Panel on Climate Change (IPCC), which we will come to a little later in this chapter, represents the unanimous view of some 2,500 scientists. In fact, the physical science section of its most recent report (IPCC, *Climate Change 2007*, February 2007) was written by 51 named authors (and subsequently edited by representatives of member governments and the UN). The other scientists engaged in the process were involved as 'reviewers' and the like, and many of these have made clear their disagreement with important aspects of the IPCC account. Then there is the even larger number of reputable climate (and allied) scientists not involved in the IPCC process, literally hundreds of whom have, at one time or another, made public their disagreement with (often fundamental) aspects of the conventional wisdom. Finally, there are large numbers of dissenting climate scientists who have chosen not to stand up and be counted, for fear that to do so would damage either their career prospects, or their chances of securing

research grants. All that can be said with confidence is that the dissenting minority of reputable climate (and allied) scientists is a sizeable one.

3. As even the distinguished earth scientist James Lovelock, originator of the imaginative 'Gaia' theory, whose view of the likely consequences of man-made global warming is at the extreme apocalyptic end of the spectrum, has written: 'I see modern science as like the mediaeval Christian church, burdened with the intricate theology of reduction. Observation and experiments are out of fashion; most evidence is now taken from the virtual world of computer models. The technique of inquisition is not the rack but the peer review: a well-intentioned instrument for sifting good from bad science that has become the great upholder of conventional wisdom.' (*Prospect,* December 2007, p. 68.) Nor of course is this defect confined to climate science: as it happens, the same issue of *Prospect* carried a brief report of a meeting at London's Royal Society to discuss synthetic biology, at which 'Several well-placed participants agreed that current funding review procedures in both Britain and the US prevent this kind of highly interdisciplinary work from thriving in the public sector, for peer review typically promotes the mediocre at the expense of the visionary and daring'. These flaws are compounded when the peers are friends and colleagues of the author, as occurred in the notorious 'hockey stick' affair.

4. See note 9.

5. K Popper, *The Logic of Scientific Discovery,* (1959).

6. Met Office/Hadley Centre, December 2005 op. cit., p. 25.

7. The 2007 Hadley Centre/CRU figure was made known to the December 2007 UN climate conference in Bali, and reported by the BBC on 13 December 2007, at http://news.bbc.co.uk/1/hi/sci/tech/7142694.stm, under the bizarre, but all too predictable heading, '2007 data confirms warming trend'. The global temperature series published by NASA's Goddard Institute for Space Studies in New York shows the same absence of further warming so far this century. It is, incidentally, generally accepted that measurements of surface temperature are at best accurate only within plus or minus 0.1°C/0.2°F.

8. D Smith et al., 'Improved Surface Temperature Prediction for the Coming Decade from a Global Climate Model', *Science*, 317, 10 August 2007, pp. 796-9.

9. See, for example, 'Will sun's low activity arrest global warming?', David Whitehouse, *The Independent,* 5 December 2007. Dr Whitehouse is an astronomer and a former online science editor of the BBC.

10. Until very recently the palm had been awarded to 1998. But the discovery (by the redoubtable Steve McIntyre – see below) of a computer bug in the treatment of the raw data led NASA's Goddard Institute for Space

Studies to issue a corrected series in August 2007. See http://data.giss.nasa.gov/gistemp/graphs/Fig.D.txt.

11. L Howard, *The Climate of London*, 1818.

12. K P Gallo et al., 'Temperature trends of the historical climatology network based on satellite-designated land use/land cover', *Journal of Climate*, 12, (1999) pp. 1344-8; E Kalnay and M Cai, 'Impact of urbanization and land-use change on climate', *Nature*, 423, (2003) pp. 528-31. The IPCC regards urbanization as only a very minor factor; but the research on which it relies for this conclusion (including the significance in the urban heat island context of wind speeds, on which the Hadley Centre rests its case: Met Office/Hadley, December 2005, op. cit. p. 27) leaves much to be desired. The Stern Review, (Nicholas Stern, *The Economics of Climate Change: The Stern Review*, 2007) while rejecting the notion that any part of the recorded 20th century warming might be due to the urban heat island effect, nevertheless warns on p. 480 that 'Climate change effects in cities are compounded by the urban heat island effect.' See, in this context (and indeed in many others), Steve McIntyre's excellent and informative blog, http://climateaudit.org.

13. It was during this phase that alarmist articles by Professor James Lovelock and a number of other scientists appeared, warning of the onset of a new ice age. See, both in this context and more generally, the very interesting written testimony given by Dr Syun-Ichi Akasofu, founding director of the International Arctic Research Centre at the University of Alaska, to the relevant Committee of the US Senate on 26 April 2006, http://iarc.uaf.edu/news/news_shorts/akasofu_4_26_06/written_testimony.php

14. See graph at: http://www.metoffice.gov.uk/research/hadleycentre/CR_data/Monthly/Hadplot_globe.gif. The precise size of each of these phases is arguable, since the Hadley Centre's own smoothed trend line is produced by a 21–year binomial filter: smoothing over a shorter period than 21 years – say, five years – would produce a rather different trend line. But the overall global temperature rise of roughly 0.7°C/1.3°F for the 20th century as a whole is not in dispute.

15. This also, incidentally, appears to be the assumption of the Stern Review. Thus it declares that 'the rising levels of [man-made] greenhouse gases provide the only plausible explanation for the observed [global temperature] trend for at least the past 50 years … the calculated warming effect is more than enough to explain the observed temperature rise', (Stern Review, op. cit., p. 8).

16. Throughout the book I shall adopt the common convention of discussing man's contribution to the greenhouse effect in terms of his contribution to carbon dioxide (CO_2) emissions and concentrations. Theoretically, it may be more correct to include other greenhouse gases to

which man's activities have contributed, and express the total in terms of the somewhat clumsy expression 'carbon dioxide equivalent', or CO_2e. But the only other greenhouse gas which is quantitatively of any significance in this context is methane (to which man has indeed contributed, largely in his role of farmer, through the belching, farting, and excretions of cows), which not only is emitted in very much smaller quantities, and remains in the atmosphere for a very much shorter time, but the considerable rise in methane concentrations during the 19th and 20th centuries appears to have come to an end (see 'Climate change and the greenhouse effect', Met Office/Hadley Centre, December 2005, p. 15; also Simpson et al., 'Influence of biomass burning during recent fluctuations in the slow growth of global tropospheric methane', *Geophysical Research Letters,* 33, issue 22, 2006). This levelling off, incidentally, was not predicted in any of the current climate models.

17. e.g. *Independent Summary for Policymakers – IPCC Fourth Assessment Report*, Ross McKitrick et al., The Fraser Institute (Canada), 2007; also P D Henderson, 'Governments and Climate Change Issues', *World Economics,* 8, no. 2, April-June 2007: a devastating indictment.

18. Fiona Harvey, 'World urged to act on definitive report', *The Financial Times*, 2 February 2007.

19. The Fourth Assessment Report of the Intergovernmental Panel on Climate Change, 2007. The Report is in three parts, the Working Group I contribution, on the science of global warming; the Working Group II contribution, on the impact of global warming; and the Working Group III contribution, on mitigation of global warming by reducing greenhouse gas emissions. Each part is in turn presented in two parts: the report itself, written by the scientists, economists, and others involved in the process, and a short 'Summary for Policy Makers' (SPM), heavily edited by the international and national bureaucrats for their political masters (and to some extent the media, although in most cases the journalists report only the press conferences given by the IPCC directorate, which are always far more alarmist than the reports on which their remarks are allegedly based). When the Third Assessment Report was published in 2001, dedicated readers discovered that there were numerous discrepancies between the full reports and the summaries, with the latter leaving out many of the important qualifications contained in the former – a fact noted, and condemned, in the House of Lords Report on the subject. This has been avoided with the Fourth Assessment Report by the expedient of publishing the (edited) summaries first, the IPCC explaining that it would ensure that by the time the full reports were published, they too would have been edited to ensure that there were no discrepancies – a somewhat unusual, not to say dubious procedure. The WGI SPM was published in February 2007, the WGII SPM in April 2007, and the WGIII SPM in May 2007. The final

part of the Fourth Assessment Report, the so-called 'Synthesis Report', was published in November 2007.

20. *Climate Change 2007: The Physical Science Basis;* Summary for Policymakers of the Contribution of Working Group I to the Fourth Assessment Report of the Intergovernmental Panel on Climate Change, 2007, p. 8. Precisely the same formulation is repeated in the Synthesis Report.

21. See, for example, 'The Stern Review: a Dual Critique – Part I: The Science', R M Carter et al., *World Economics,* 7 no. 4, October-December 2006.

22. R Lindzen et al., 'Does the Earth have an Adaptive Infrared Iris?', *Bulletin of the American Meteorological Society*, 82, no. 3, (2001) pp. 417-32. Richard Lindzen is Professor of Atmospheric Physics at the Massachusetts Institute of Technology (MIT).

23. IPCC, 2007, op. cit., WGI, SPM p 9.

24. Met Office/Hadley Centre, 2005, op. cit., p 22.

25. Met Office/Hadley Centre, 2005, op. cit., p 25.

26. See T L Anderson, et al., 'Climate forcing by aerosols – a hazy picture', *Science,* 300, (May 2003), pp. 1103-04. It is also worth noting that the most recent observational evidence, derived from measurements of the Asian 'brown cloud', suggests that aerosols are actually having a *warming* effect over a large part of Asia. If this finding is confirmed, it will of course seriously undermine the conventional scientific wisdom. See V Ramanathan et al., 'Warming trends in Asia amplified by brown cloud solar absorption', *Nature*, 448, 2007, pp. 575-8.

27. Met Office/Hadley Centre, 2005, op. cit., p. 29.

28. Met Office/Hadley Centre, 2005, op. cit., p. 10.

29. Met Office/Hadley Centre, 2005, op. cit., p. 32.

30. Another discrepancy concerns rainfall, which the models claim to predict. A recent study has found that, since 1987, the actual increase in global rainfall has been up to seven times as great as that predicted by the climate models (F Wentz et al., 'How much more rain will global warming bring?', *Science*, 317, 2007, pp. 233-5).

31. K Trenberth, 'Predictions of Climate', *Climate Feedback: The climate change blog*, (June 2007) http://blogs.nature.com/climatefeedback/recent_contributors/kevin_trenberth/ .

32. The fact that the sun has a major impact on the world's climate is of course blindingly obvious. Whether its activities have been more important than man-made enhancement of the greenhouse effect in accounting for the modest 20th century global warming is a matter of lively debate. It is not just a matter of changes in solar radiation, but also of the sun's influence on cosmic rays, and the effect of cosmic rays on clouds. The Hadley Centre reports (op.

cit. p. 10) that a study which it commissioned 'concluded [in 2005] that there is some empirical evidence for relationships between solar changes and climate, and several mechanisms, such as cosmic rays influencing cloudiness, have been proposed, which could explain such correlations. These mechanisms are not sufficiently well understood and developed to be included in climate models at present.' The leading proponent of the cosmic ray hypothesis as the explanation of 20th century global warming is Dr Henrik Svensmark, director of the Centre for Sun-Climate Research at the Danish National Space Centre. For an extended discussion of this, see *The Chilling Stars: A New Theory of Climate Change,* Henrik Svensmark & Nigel Calder, 2007; see also, inter alia, J Lean et al., 'Reconstruction of solar irradiance since 1610: implications for climate change', *Geophysical Research Letters*, 22, 1995, pp. 3195-8; D V Hoyt and K H Schatten, 'A discussion of plausible solar irradiance variations, 1700-1992', *Journal of Geophysical Research*, 98, 1993, pp. 18895-906; S K Solanki and M Fligge, 'Solar irradiance since 1874 revisited', *Geophysical Research Letters*, 25, 1998, pp. 341-4; and H Svensmark et al., 'Experimental evidence for the role of ions in particle nucleation under atmospheric conditions', *Proceedings of the Royal Society A*, October 2006.

33. See, for example, Huang et al., 'Late Quaternary temperature changes seen in world-wide continental heat flow measurements', *Geophysical Research Letters*, 24, issue 15, 1997, pp. 1947-50; G Bond et al., 'Persistent solar influence on North Atlantic climate during the Holocene', *Science,* 294, 2001, pp. 2130-6; and L Keigwin, 'The Little Ice Age and Mediaeval Warm Period in the Sargasso Sea', *Science,* 274, 1996, pp. 1504-08. See also H Lamb, *Climate, History and the Modern World,* (1995).

34. A G Brown et al., 'Roman vineyards in Britain: stratigraphic and palynological data from Wollaston in the Nene Valley, England', *Antiquity,* 75, no. 290, 2001, pp. 745-57. And as Lamb has pointed out, 'The cultivation of grapes for wine making was extensive throughout the southern portion of England from about 1100 to around 1300' (H Lamb, 'The early Mediaeval warm epoch and its sequel', *Paleogeography, Paleoclimatology, Paleoecology*, 1, 1965, pp. 13–37).

35. See for example C I Millar, J C King, R D Westfall, H A Alden and D L Delany, 'Late Holocene forest dynamics, volcanism, and climate change at Whitewing Mountain and San Joaquin Ridge, Mono County, Sierra Nevada, CA, USA', *Quaternary Research*, 66, 2006, pp. 273-87; Mukhtar M Naurzbaev, Malcolm K Hughes, Eugene A Vaganov, 'Tree-ring growth curves as sources of climatic information', *Quaternary Research*, 62, 2004, 126-33.

36. A Newton, R Thunell, and L Stott, 'Climate and hydrographic

variability in the Indo-Pacific Warm Pool during the last millennium', *Geophysical Research Letters*, 33, 2006; J N Richey, R Z Poore, B P Flower, T M Quinn, '1400 year multiproxy record of climate variability from the northern Gulf of Mexico', *Geology*, 35, 2007, pp. 423–26; B K Khim, H I Yoon, C Y Yang and J J Bahk, 'Unstable climate oscillations during the late Holocene in the Eastern Bransfield Basin, Antarctic Peninsula', *Quaternary Research*, 58, 2002, pp. 234–45.

37. M Mann et al., 'Global-scale temperature patterns and climate forcing over the past six centuries', *Nature,* 392, 1998, pp. 779–87; also M Mann et al., 'Northern hemisphere temperatures during the past millennium: inferences, uncertainties, and limitations', *Geophysical Research Letters,* 26, 1999, pp. 759–62.

38. The email reply to Warwick Hughes dated 21 February 2005 was reported on 2 March 2006 in slide 4 of the presentation Professor Hans von Storch (of the Meteorological Institute of Hamburg University) gave to the US National Research Council Committee on Surface Temperature Reconstruction for the Last 2,000 Years. Dr von Storch reported that he had personally verified the email with Dr Jones. Some of the withheld data was subsequently released by the CRU in response to Freedom of Information Act requests in October 2007.

39. It turned out, inter alia, that crucial data for the period before 1421 was based on an analysis of tree rings in one single pine tree. (When this came to light, it was – inevitably – dubbed 'the lonesome pine'.) For an informed discussion of the issue, see S McIntyre and R McKitrick, 'Corrections to the Mann et al. (1998) proxy data base and northern hemisphere average temperature series', *Energy and Environment*, 14, no. 6, November 2003, pp. 751-71; S McIntyre and R McKitrick, 'Hockey Sticks, Principal Components, and Spurious Significance', *Geophysical Research Letters*, 32, 2005; and E J Wegman et al., 'Ad Hoc Committee Report on the "Hockey Stick" Global Climate Reconstruction', (July 2006), which concluded, inter alia, that 'It is important to note the isolation of the palaeoclimate community; even though they rely heavily on statistical methods they do not seem to be interacting with the statistical community. In addition, we judge that there was too much reliance on peer review, which was not necessarily independent ... Overall our committee believes that Dr Mann's assessments that the decade of the 1990s was the hottest decade of the millennium and that 1998 was the hottest year of the millennium cannot be supported by his analysis.' For an excellent non-technical account of the 'hockey stick' saga, see D Holland, 'Bias and concealment in the IPCC process: the "Hockey-Stick" affair and its implications', *Energy and Environment*, 18 (7-8), 2007; also R M Carter et

al., 'The Stern Review: a Dual Critique – Part I: The Science', *World Economics,* 7, no. 4, October-December 2006.

40. G D North, 'Surface Temperature Reconstructions for the Last 2,000 Years'. 'Statement before the Subcommittee on Oversight and Investigations Committee on Energy and Commerce', *US House of Representatives,* 19 July 2006, at http://dels.nas.edu/dels/rpt_briefs/Surface_Temps_final.pdf.

41. In his Review Comments on the report Dr Mann describes this omission as 'effectively a slap in the face'.

42. J Hirschi et al., 'Global warming and changes of continentality since 1948', *Weather*, 82, no. 8, August 2007.

43. See, for example, C Vincent et al., 'Very high-elevation Mont Blanc glaciated areas not affected by the 20th century climate change', *Journal of Physical Research*, 112, 2007; also V Raina and C Sangewar, 'The Siachen glacier', *Journal of the Geological Society of India*, 70, no. 1, July 2007, pp. 11-16.

44. H Zwally et al., 'Mass changes of the Greenland & Antarctic ice sheets & shelves & contributions to sea-level rise 1992–2002', *Journal of Glaciology*, 51, no. 175, December 2005, pp. 509-27; O Johannessen et al., 'Recent Ice-Sheet Growth in the Interior of Greenland', *Sciencexpress*, 20 October 2005; W Krabill et al., 'Greenland ice sheet: high-elevation balance and peripheral thinning', *Science*, 289, 2005, pp. 428-30.

45. See, for example, N Morner et al., 'New perspectives for the future of the Maldives', *Global and Planetary Change,* 40, issues 1-2, January 2004, pp. 177-82; and, for the wider complexities of interpreting sea level rise and fall, N Morner, 'Estimating future sea level changes from past records', *Global and Planetary Change*, 40, issues 1-2, January 2004, pp. 49-54. See also S Jevrejeva et al., 'Nonlinear trends and multiyear cycles in sea level records', *Journal of Geophysical Research*, 111, 2006.

46. N Stern, *The Economics of Climate Change: The Stern Review,* (2007), p 10.

47. This is conceded in the body of IPCC reports, but played down or ignored in the summaries. Buried on page 505 of the IPPC Third Assessment Working Group I Report (2001) was the admission that 'In climate research and modelling, we should recognize that we are dealing with a coupled non-linear chaotic system, and therefore that the long-term prediction of future climate states is not possible.' The latest report is more evasive but, for example, on aerosols, still concedes that 'The large uncertainty in total aerosol forcing makes it more difficult to accurately infer the climate sensitivity from observations' (section 9.6). It also increases uncertainties in results that attribute cause to observed climate change (section 9.4.1.4), and is in part responsible for differences in probabilistic projections of future climate

change (chapter 10). 'Forcings from black carbon, fossil fuel organic matter and biomass burning aerosols, which have not been considered in most detection studies performed to date, are likely small but with large uncertainties relative to the magnitudes of the forcings.' (IPCC Fourth Assessment Working Group I Report, pages 678-9). On clouds and precipitation, page 592 of the same report gingerly notes that 'In some models, simulation of marine low level clouds, which are important for correctly simulating sea surface temperature and cloud feedback in a changing climate, has also improved. Nevertheless, important deficiencies remain in the simulation of clouds and tropical precipitation (with their important regional and global impacts).'

48. The most meritorious sentence in the review (and the least publicized) is the last: 'The analysis of the Review as a whole was always intended to be one contribution to a discussion. There have been, will be, and should be many more contributions.' Among the 'many more contributions' see, for example, Byatt, Castles, Goklany, Henderson, Lawson, McKitrick, Morris, Peacock, Robinson and Skidelsky, 'The Stern Review: a dual critique; Economic aspects', *World Economics*, 7, (4), 2006, pp. 199-29, including I C Byatt et al., 'Part II: The economics'.

49. The 50-page dossier entitled 'Iraq's Weapons of Mass Destruction: The Assessment of the British Government' was published on 24 September 2002 in order to enable the then prime minister, Mr Blair, to secure parliamentary support for the invasion of Iraq. In his foreword Mr Blair declared, inter alia, that 'Saddam Hussein is continuing to develop WMD … and the document discloses that his military planning allows for some of the WMD to be ready within 45 minutes of an order to use them'. The document, which also contained a number of other unfounded alarmist assertions, subsequently became widely known in the UK as Mr Blair's dodgy dossier.

Chapter 2: The Next Hundred Years: How Warm? How Bad?

1. Even the ultra-gloomy Professor Lovelock observes that 'earth history suggests that positive feedback will come to a natural stop and temperatures will stabilize five degrees above the present'. (Lovelock J, *Prospect,* December 2007).

2. Evidence to House of Lords Economic Affairs Committee, 18 January 2005, HL Paper 12-II, p. 23.

3. IPCC, Third Assessment Report, 2001, Technical Summary of the Working Group I report, figure 28, p. 79.

4. N Nakicenovic et al., *Special Report on Emissions Scenarios*, 2000.

5. This crass error was pointed out as far back as 2003 by Ian Castles, former

head of the Australian Bureau of Statistics, and Professor David Henderson, former head of the Economics and Statistics Department of the OECD. See I Castles and D Henderson, 'The IPCC emissions scenarios: and economic-statistical critique', *Energy and Environment*, 14: 2&3, 2003, pp. 159-86.

6. K Trenberth, 'Global Warming and Forecasts of Climate Change', *Climate Feedback: The climate change blog*, (July 2007) http://blogs.nature.com/climatefeedback/recent_contributors/kevin_trenberth/. This point is lost on the lamentable Stern Review, incidentally, which routinely uses the term 'prediction' and confidently tells the reader what 'will' happen.

7. IPCC, Fourth Assessment Report, 2007, op. cit., table SPM-3, p. 13. These projected increases are in fact changes from the 1980-1999 average temperature. To express the change relative to the current estimated world average temperature it is necessary to deduct 0.2ºC/0.4ºF, i.e. the 'best estimate' range is an increase of between 1.6ºC/2.9ºF and 3.8ºC/6.8ºF compared with the present. The IPCC's 'best estimates' represent its best estimates within a wider range of possible temperature rises, of between 1.1ºC/2.0ºF and 6.4ºC/11.5ºF compared with 1980-99 (or between 0.9ºC/1.6ºF and 6.2ºC/11.2ºF compared with the present).

By contrast, the Stern Review, published before the IPCC's Fourth Assessment Report appeared, let alone the acknowledgment of the 21st century warming standstill, takes as its base case a rise in global temperature of between 5ºC/9ºF and 6ºC/10.8ºF by some time in the 22nd century. Stern's rise, incidentally, is explicitly compared with 1750-1850 temperatures, a period before global temperature records existed. But if we take his base to mean 1850-1899 levels, preferred by the IPCC as the earliest for which records exist, this would imply a base case rise of some 4.3-5.3ºC/7.7–9.5ºF above today's levels. To get its consequent damage estimates, the review then feeds this into one particular and carefully chosen computer model, known as PAGE2002, which, according to the review, is based on 'choosing a set of uncertain parameters from pre-determined ranges of possible values'. It then substantially inflates, on various dubious and highly subjective grounds, the estimated damages that emerge from this exercise.

8. I M Goklany, *The Improving State of Humanity*, 2007. Dr Goklany is Assistant Director, Science & Technology Policy, Office of Policy Analysis, US Department of the Interior, and is a former US delegate to the IPCC.

9. Another form of reality check is to look at the stock market. While business leaders increasingly feel it is good PR to parade their concern at global warming, stock market prices in general do not reflect any forthcoming global disaster or even any risk of it. Nor, for that matter, do carbon-based energy stocks reflect any expectation of the recommended global decarbonization actually occurring. There has, of course, been stock market

interest in renewable energy, not because of any expected major role in overall energy supplies, but reflecting the huge subsidies being thrown at it, and also in those firms engaged in the potentially highly profitable business of carbon trading (discussed in Chapter 6).

10. The account that follows is taken directly from *Climate Change 2007: Climate Change Impacts, Adaptation and Vulnerability*; Summary for Policymakers of the Working Group II Contribution to the Fourth Assessment Report of the Intergovernmental Panel on Climate Change, 2007.

11. IPCC, 2007, op. cit., table SPM-1.

12. For instance table SPM 2 of the IPCC's Synthesis Report, (IPCC, 2007, op. cit.) stretching over two pages, is entitled 'Examples of some projected regional impacts', and gives four examples for each of the eight regions into which it divides the world. Each and every one of the 32 impacts listed is adverse.

13. See, in this context, Gedney et al., 'Detection of a direct carbon dioxide effect in continental river runoff records', *Nature*, 439, 16 February 2006, pp. 835-8; a Hadley Centre study in which it is suggested that this might be due to the fertilization effect of the increase in CO_2 concentrations in the atmosphere, since the greater efficiency of photosynthesis caused by the increased CO_2 means that plants need to use less water in order to thrive.

14. For a fuller treatment of this, see K Okonski (ed.), *The Water Revolution: Practical Solutions to Water Scarcity*, 2006.

15. See, for example, I M Goklany, 'Saving habitat and conserving biodiversity on a crowded planet', *BioScience*, 48, 1998, pp. 941-53. Also 'CO_2 and biodiversity: Does the former affect the latter?', *CO_2 Science*, 5, no. 35, August 2002.

16. See D Botkin, 'Global Warming Delusions', *Wall Street Journal,* 17 October 2007. Daniel Botkin is President of the Center for the Study of the Environment and professor emeritus in the Department of Ecology, Evolution and Marine Biology at the University of California, Santa Barbara.

17. D Darby et al. ('New record shows pronounced changes in Arctic Ocean circulation and climate', *EOS, Transactions,* 82, no. 49, 2001, pp. 601-7), suggested that average temperatures in the Arctic have varied by 6°C/10.8°F over the past 8,000 years. G M MacDonald et al. ('Holocene treeline history and climate change across northern Eurasia', *Quaternary Research*, 53, 2000, pp. 302-11), concludes that in the Eurasian Arctic, temperatures were 2.5-7°C/4.5–12.6°F warmer at the peak of the present interglacial period (7,000 to 9,000 years ago) than they are now. The IPCC Fourth Assessment Working Group I Report (op. cit., p. 462) lists numerous studies that have indicated significantly warmer temperatures in the northern high latitudes at various points between 4,000 and 10,000 years ago; the Nordic Seas being more than 2°C/3.6°F warmer than recent

temperatures between 7,500 and 10,000 years ago. CAPE Project Members ('Last interglacial Arctic warmth confirms polar amplification of climate change', *Quaternary Science Reviews*, 25, 2006, pp. 1383-1400), claim that the entire Arctic was also warmer than it is now during the last interglacial about 125,000 years ago, much of it by 4ºC/7.2ºF or more.

18. On polar bears, see also B Lomborg, *Cool It: The Skeptical Environmentalist's Guide to Global Warming,* 2007, pp. 4-7. This book is well worth reading in its entirety.

19. S J Holgate, 'On the decadal rates of sea level change during the twentieth century', *Geophysical Research Letters*, 34, no. 1, January 2007.

20. See note 45, p. 114 above.

21. See V Gray, 'The truth about Tuvalu', *NZ Climate and Enviro Truth,* 103. Tuvalu, and the other small islands of the region, are coral atolls; and it has been suggested that, as and when sea levels in the region rise, the coral rises, and with it the atolls, as they are constantly built up with coral sand moved by wind and wave from the reef. Hence the measured stability. See W Eschenbach, 'Sinking in Tuvalu', *Energy and Environment,* 15(3), 2004.

22. IPCC, 2007, op. cit., 'The Physical Science Basis', table SPM-2. It is interesting, incidentally, that the IPCC's estimate of the upper limit of the possible rise in the sea level by 2100 has been steadily coming down in each successive assessment report in the light of new evidence and new studies. Thus the first (1990) report put it at 367 centimetres, the second (1995) report at 124 centimetres, the third (2001) report at 77 centimetres, and now the fourth assessment report puts it at 59 centimetres.

23. P Reiter, 'Malaria in England in the Little Ice Age', *Emerging Infectious Diseases*, 6, no. 1, January-February 2000. See also Professor Reiter's written evidence to the House of Lords Select Committee, HL Paper 12-II, pp. 284-8.

24. See R Tren and R Bate, *Malaria and the DDT Story*, 2001. The malaria scare is among the many featured in the film, *An Inconvenient Truth*, in which Mr Gore claims that the Kenyan capital, Nairobi, was 'originally located just above the mosquito line', but that 'now, with global warming, the mosquitoes are climbing to higher altitudes'. The really inconvenient (for Mr Gore) truth is that Nairobi was malaria-ridden from the start. The city's first medical officer, a Dr Boedeker, recorded that it 'had always been regarded as an unhealthy locality swarming with mosquitoes' and that, in 1904, a committee of doctors 'petitioned that the entire municipality be relocated, simply because it was a spawning ground of disease'. A substantial reduction in the incidence of malaria occurred during the 1950s, with the introduction of DDT; and although there has been some resurgence over the past 20 years, largely

due to the effective ban on DDT, it is still well below the levels of the earl
years.

25. IPCC 2007, op. cit., SPM, WGII, pp. 9–10.

26. HL Paper 12-II of 2005, p. 20. The Stern Review (Stern, 2007, op. cit.) makes the same point, no fewer than three times (in the executive summary and on both pages 122 and 133 of the main report), although it ups the number of deaths to 35,000.

27. See W Keatinge, 'Seasonal mortality among elderly people with unrestricted home heating', *British Medical Journal*, 293, 1986, pp. 732-3, and W Keatinge et al., 'Heat related mortality in warm and cold regions of Europe: observational study', *British Medical Journal*, 321, 2000, pp. 670-3.

28. W Keatinge and G Donaldson, 'The impact of global warming on health and mortality', *Southern Medical Journal*, 97, no. 11, 2004, p. 1096. See also M Laaidi et al., 'Temperature-related mortality in France; a comparison between regions with different climates from the perspective of global warming', *International Journal of Biometeorology*, 51, 2006, pp. 145-53, where there is a similar finding.

29. Table 3.2, which is described as 'Examples of possible impacts of climate change due to changes in extreme weather and climate events, based on projections to the mid- to late-21st century'. The bulk of them, needless to say, are adverse impacts. Characteristically (see Chapter 3 for discussion of this), the IPCC adds that 'These do not take into account any changes or developments in adaptive capacity.'

30. IPCC 2007, op. cit., WGII, SPM, p. 20.

31. See Nakicenovic, 2007, op. cit., and http://www.unpopulation.org

32. There is, in fact, an extensive academic and public policy literature on all this. For a convenient summary of much of it, see for example the recent report by the House of Lords Select Committee on Economic Affairs, 'Government Policy on the Management of Risk', HL Paper 183-1, June 2006.

Chapter 3: The Importance of Adaptation

1. It is true that the IPCC's Synthesis Report does state (Topic 5, p. 3, IPCC, 2007, op. cit.) that 'Adaptation and mitigation can complement each other and together can significantly reduce the risks of climate change.' The two approaches are treated in Topics 4 and 5 of the Synthesis Report. The 9-page long Topic 4 devotes two somewhat dismissive pages to adaptation and seven enthusiastic pages to mitigation. The treatment in the 10-page long Topic 5 is even more unbalanced. This imbalance appears to be justified by the absurd assertion that 'much less information is available about the costs and effectiveness of adaptation measures than about mitigation levels'. To the

extent that this statement is true, it is simply because the IPCC has made no attempt to assess the cost, and precious little attempt to assess the effectiveness, of adaptation.

2. N Stern, 2007, op. cit., p. 97 and footnote on p. 119.

3. Bangladesh has, in effect, experienced a slightly falling sea level over the past hundred years or so, as a result of silt deposition. But the monsoon flooding damage has worsened, at least in part because many of the canals, constructed when the country was part of Britain's Indian Empire to allow the flood waters to escape, have been filled in as part of the Bangladesh government's road-building programme.

4. There does, regrettably, appear to be a somewhat collectivist and anti-market bias to the economic sections of the IPCC Report. Thus the only form of adaptation it is prepared to entertain is what it describes as 'planned adaptation', and it claims that 'vulnerability to climate change can be exacerbated by other stresses', such as 'trends in economic globalization'. Needless to say, the reverse is the case.

5. N Stern, 'What is the Economics of Climate Change?', *World Economics*, 7, no. 2, April-June 2006.

6. F A Hayek, *The Economy, Science, and Politics*, Routledge & Kegan Paul, 1967.

7. Even 'thus far but no further' would, incidentally, require *inter alia* rigid global population control. The threefold growth of world population during my own lifetime, from about 2 billion when I was born to around 6 billion now, has itself contributed significantly to the growth of unequivocally man-made CO_2 emissions. (We exhale carbon dioxide every time we breathe.)

Chapter 4: Apocalypse and Armageddon

1. N Stern, 2007, op. cit., p. 331.

2. See also, in this context, R Pielke Jr, 'Mistreatment of the economic impacts of extreme events in the Stern Review', *Global Environmental Change*, 2007.

3. See e.g. N Stern, 2007, op. cit., p. 21, footnote 66.

4. M Hulme, 'Chaotic world of climate truth', BBC News, 4 November 2006, available from;
http://news.bbc.co.uk/go/pr/fr/-/2/hi/science/nature/6115644.stm

5. 'Sense about Science', *Making Sense of the Weather and Climate*, 2007.

6. R Pielke and C Landsea, 'Damage trends in Atlantic hurricanes', *Natural Hazards Review*, 2007. See also 'The deadliest hurricanes in the United States 1900-1996', National Hurricane Center, at http://www.nhc.noaa.gov/pastdead.html, and the Met Office news release 'Tropical storms

and climate change', 20 February 2006, at; http://www.metoffice.com/corporate/pressoffice/2006/pr20060220.html.

7. 'Consensus statements by International Workshop on Tropical Cyclones-VI participants'(IWTC-VI), contributed by Chris Landsea. Hurricane Research Division FAQ, Atlantic Oceanographic and Meteorological Laboratory; http://www.aoml.noaa.gov/hrd/tcfaq/G3.html. For the full 'Statement on Tropical Cyclones and Climate Change' see http://www.wmo.ch/pages/prog/arep/tmrp/documents/iwtc_statement.pdf.

8. B M Vinther et al., 'Extending Greenland temperature records into the late eighteenth century', *Journal of Geophysical Research*, 3, 2006.

9. D Wingham et al., 'Mass balance of the Antarctic ice sheet', *Philosophical Transactions of the Royal Society*, A, 364, 2006, pp 1627-35.

10. IPCC, 2007, op. cit., WGII, SPM, p. 17. It will be recalled that 'medium confidence' is defined by the IPCC as meaning that the chances of the prediction being correct are 'about 5 out of 10' – for what that is worth in this context.

11. Met Office/Hadley Centre, December 2005 op. cit., p. 9.

12. F Schott et al., *Geophysical Research Letters*, 33, 2006. See also S Cunningham et al., 'Temporal variability of the Atlantic Meridional Overturning Circulation at 26.5ºN', *Science,* 317, August 2007, pp. 935-8.

13. See C Wunsch, 'Gulf Stream safe if wind blows and Earth turns', *Nature*, 428, 2004. See also his letter in *The Economist*, 30 September 2006. Carl Wunsch is Professor of Physical Oceanography at the Massachusetts Institute of Technology (MIT).

14. IPCC, 2007, op. cit., WGI, SPM, p. 12.

15. S Seager et al., 'Is the Gulf Stream responsible for Europe's mild winters?', *Quarterly Journal of the Royal Meteorological Society*, 128, no. 586, 2002.

Chapter 5: A Global Agreement?

1. 'China climate plan stresses right to fast growth', *Deutsche Welle,* 4 June 2007; Ma Kai, 'China is shouldering its climate change burden', *Financial Times,* 4 June 2007.

2. Anita Chang, 'China answers emissions critics', *Associated Press*, 22 June 2007.

3. D Helm et al., 'Too good to be true? The UK's Climate Change Record', available at http://www.dieterhelm.co.uk/publications/ Carbon_record_2007.pdf.

4. Richard Spencer & Peter Foster, 'China and India reject climate change deal', *The Daily Telegraph,* 9 June 2007.

5. N Stern, 2007, op. cit., p. 535.

6. Andrew McCathie, 'Roundup: China, India insists climate change solution lies in west', *Deutsche Presse Agentur*, 8 June 2007.

7. For the latest emissions data see: **USA**: Environmental Protection Agency, *US Inventory of Greenhouse Gas Emissions and Sinks 1990-2005*, (2007), available at http://www.epa.gov/climatechange/emissions/downloads/2007GHGFastFacts.pdf. **Europe**: European Environment Agency, *Annual European Community greenhouse gas inventory 1990-2005 and inventory report*, Submission to the UNFCCC Secretariat. EEA Technical Report, No. 7, (2007). Available at http://reports.eea.europa.eu/technical_report_2007_7/en/Annual%20European%20Community%20greenhouse%20gas%20inventory%201990-2005%20and%20inventory%20report%202007.pdf. **Canada**: Greenhouse Gas Division of Environment Canada, *National Inventory Report, 1990-2005: Greenhouse Gas Sources and Sinks in Canada*, The Canadian Government's Submission to the UN Framework Convention on Climate Change, (April 2007). Available at http://www.ec.gc.ca/pdb/ghg/inventory_report/2005_report/tdm-toc_eng.cfm

8. The Bali communiqué acknowledged the 'common but differentiated responsibilities and respective capabilities' of different countries at different stages of development, and called for 'nationally appropriate mitigation actions by developing country parties'. The prominent Green campaigner George Monbiot described it as 'even worse than the Kyoto protocol' (*The Guardian,* 17 December 2007). As for post-Bush America, Senator Kerry, the defeated Democrat candidate last time round, and a committed Goreite, declared in Bali that, if China and other emerging economies did not commit to greenhouse gas reductions, it would be 'very difficult' to get a new global climate deal through the US Senate, even under a Democratic President.

9. 'Draft options paper on renewables target', Department for Business, Enterprise and Regulatory Reform, leaked to *The Guardian*, August 2007.

10. Another example of EU inconsistency in this area concerns two decisions taken in 2007, both of which are scheduled to come into force in 2009. The first was to ban the sale of mercury thermometers, on the grounds that they are a health hazard, since mercury is a poison. The second was to ban the sale of traditional (incandescent) light bulbs in order to force residents of the European Union to use energy-saving (fluorescent) light bulbs – despite the fact that the latter, as well as being distinctly unattractive, also contain mercury, and thus constitute a far greater health hazard than the very much smaller number of mercury thermometers.

Chapter 6: The Cost of Mitigation

1. Met Office/Hadley Centre, December 2005 op. cit., p. 17.
2. HL paper 12-II, 2005, p. 93.

3. *World Energy Outlook 2007*: 'The Reference Scenario', International Energy Agency, p. 74.

4. A further complication is that as technological development increases energy efficiency, this has the effect of lowering the cost of energy to the user, thus reducing the extent to which he or she will use less of it. Thus, for example, if technological development enables motorcars to do more miles per gallon, some users may feel that they can afford to drive more miles, or indeed trade up to a larger car.

5. IPCC, 2007, op. cit., WGIII, SPM, p. 31.

6. OECD, 'Biofuels: Is the cure worse than the disease?', Round Table on Sustainable Development, September 2007.

7. At the March 2007 European Council it was agreed that a minimum of 10% of EU transport petrol and diesel consumption should come from biofuels by 2020.

8. See for example D Pimentel, 'Ethanol fuels: Energy balance, economics and environmental impacts are negative', *Natural Resources Research*, 12, no. 2, June 2003. If forests are cleared to grow biofuel crops, the net effect on atmospheric CO_2 is clearly adverse: see for example R Righelato & D Spracklen, 'Carbon mitigation by biofuels or by saving and restoring forests?', *Science*, 317, no. 5840, August 2007, p. 902.

9. OECD, op. cit., September 2007. The Report notes, inter alia, that the cost of obtaining a unit of CO_2-equivalent reduction through subsidies to biofuels is well over $500 per tonne of CO_2-equivalent in the US, and substantially more than that in the EU.

10. Department for Trade and Industry, Energy White Paper, 2003, 'Our Energy Future – Creating a Low Carbon Economy'.

11. See, for example, D Simpson, 'Tilting at Windmills: the economics of wind power', David Hume Institute, *Occasional Paper no 65*, April 2004. David Simpson, former Professor of Economics at Strathclyde University, concludes that 'the cost of generating electricity from wind power is twice that of the cheapest alternative conventional source'. See also M Laughton, 'Observations on the UKERC Report on *The Costs and Impacts of Intermittency*', April 2006 (Michael Laughton is Emeritus Professor of Engineering at London University), and also Malcolm Keay, *The Dynamics of Power: Power Generation Investment in Liberalised Energy Markets*, 2006, where, in Chapter 6, it is argued that, largely because of this problem, 'the most likely outcomes are ones in which the introduction of renewables leads to little, if any, CO_2 saving'. Finally, see also Tyndall Centre, *Security Assessment of Future UK Electricity Scenarios*, Conclusions, pp. 47-8.

12. Prime Minister Gordon Brown's foreword to the Government's *White*

Paper on Nuclear Power, Department for Business, Enterprise and Regulatory Reform, January 2008.

13. International Energy Agency, *World Energy Outlook, 2007*; Table 1.1, 'World Primary Energy Demand in the Reference Scenario', p. 74.

14. Jeroen Van de Veer, 'High hopes and hard truths dictate future', *The Times*, 25 June 2007. The present Chancellor of the Exchequer, Mr Alastair Darling, when he was the minister responsible for energy policy, warned the House of Commons that the technologies required for commercial carbon capture and storage 'might never become available' (*Hansard*, 23 May 2007, col. 1289 available at http://www.publications.parliament.uk/pa/ld200405/ldhansrd/vo050301/text/50301-12.htm).

15. Adam Smith, *An Inquiry into the Nature and Causes of the Wealth of Nations*, 1976.

16. Another reason why the threat from dependency on Russian gas is confined to the admittedly acute problems of a purely temporary interruption of supplies is that it is in principle a relatively straightforward matter to retrofit a modern gas-fired power station with a gasifier to enable it to run on synthetic gas ('syngas') made from coal.

17. Margaret Thatcher alludes to this, very briefly, in her memoirs: *The Downing Street Years*, 1993, pp. 140 and 640.

18. For a fuller discussion of this issue see D Howell and C Nakhle, *Out of the Energy Labyrinth*, 2007.

19. See, for example, 'Europe's dirty secret: Why the EU Emissions Trading Scheme isn't working', available from http//:www.openeurope.org.uk (2007). See also Ed Crooks, *Financial Times*, 17 June 2007.

20. In January 2008, the European Commission proposed that, for the third stage of the ETS, scheduled to start in 2013, at least two-thirds of all permits should be auctioned. At the time of writing, this seems unlikely to be accepted by the EU member states.

21. See, for example, 'Abuse and incompetence in fight against global warming' and 'Truth about Kyoto: huge profits, little carbon saved', both by Nick Davies in *The Guardian*, 2 June 2007.

22. George Monbiot, writing in *The Guardian*, 17 December 2007.

23. Roger Harrabin, 'Rich "can pay poor to cut carbon"', *BBC News*, 22 August 2007 available at http://news.bbc.co.uk/1/hi/sci/tech/6957328.stm.

24. See, for example, Fiona Harvey et. al., 'Producers, traders reap credits windfall', and 'Defra in storm over EU carbon scheme', in *The Financial Times*. Also Nick Davies in *The Guardian*, 'The inconvenient truth about the carbon offset industry'.

25. IPCC, WGIII SPM Second Order Draft, July 2006, p. 8. It notes that 'These results are from top-down models that assume least cost application of

mitigation in all regions, but do not make assumptions on who should pay for this mitigation'. The second order drafts are not published documents, but they are circulated widely both to member governments and to those involved in the IPCC process.

26. IPCC, 2007, op. cit., WGIII, SPM, p. 27.

27. Ibid. Some, incidentally, have sought to trivialize the cost of mitigation by arguing that, assuming the world economy grows at some 2.5% a year, if mitigation costs 2.5% of world GDP that simply means that the world economy will be where it would have been a year later, and even if the cost is 5%, that merely means a two-year delay. But precisely the same formulation can be used to trivialize the damage that might be caused by global warming; if nothing whatever is done about it the cost will simply be a delay of a year or two in the onward march of economic growth. In fact, in both contexts, this attempted trivialization is equally misleading.

28. IPCC, 2007, op. cit., WGIII, SPM, pp. 27-8.

29. See, for example, Bjorn Lomborg (ed.), *Global Crises, Global Solutions: Priorities for a world of scarcity*, 2004; also http://www. copenhagenconsensus. com. See also Lomborg's testimony ('Perspective on Climate Change') to the US Senate's joint hearing of the Committee on Energy and Environment and the Committee on Science and Technology, 21 March 2007.

30. See N Lawson, *The View from No. 11*, 1992, pp. 739-44.

Chapter 7: Discounting the Future: Ethics, Risk and Uncertainty

1. IPCC, 4AR WGIII Second Order Draft, p. 8.

2. According to Professor Weitzman (see note 4, p. 126) Stern assumes a discount rate of only 1.4%. Given the critical importance of the discount rate chosen, it is surprising, to say the least, that nowhere in the 692-page Stern Review does its author make explicit precisely what discount rate does lie at the heart of its economic analysis.

The review's treatment of all this is to be found in its Chapter 2A, 'Ethical Frameworks and Intertemporal Equity'. As is customary in welfare economics, this is littered with mathematical equations and garnished with letters of the Greek alphabet; but what matters is the values attached to the so-called ethical parameters represented by the Greek letters, delta and eta, which is inevitably wholly subjective. The framework is the conventional one derived from Frank Ramsey's pioneering work, 'A mathematical theory of saving', *Economic Journal*, 38, 1928, pp. 543-59. It leaves much to be desired, not least the assimilation of risk-aversion and inequality-aversion in a single variable (see W Beckerman and C Hepburn, 'Ethics of the Discount Rate in the Stern Review on the Economics of Climate Change', *World Economics*, 8, no. 1, January-March 2007) and it is hard to believe that had

Ramsey, who died in 1930 at the age of 26, lived long enough to return to the subject, he would not have been able to refine and improve this framework. (The cleverest member of a gifted family – his father was President of Magdalene College, Cambridge, and his younger brother became Archbishop of Canterbury – Ramsey was widely regarded by his contemporaries at Cambridge, who included Keynes and Wittgenstein, as having the finest mind of his generation. A mathematician by trade, during his short life he made a number of major original contributions to mathematics, philosophy – his principal interest – and economic theory, most of which remain of importance today.)

In essence, Stern (following Ramsey) holds that the relevant discount rate is a combination (i.e. addition) of the pure rate of time preference (conventionally signified by the Greek letter delta) – that is, the discount attached to the welfare of future generations compared with that of the present generation – and the assumed rate of growth of consumption per head (which, in the long run, is equivalent to the assumed rate of growth of GDP per head) multiplied by the so-called 'elasticity of marginal utility' (conventionally signified by the Greek letter eta). This last is essentially a measure of how much a loss (or gain) in income for a poor man affects his welfare, compared with how much the same percentage loss (or gain) in income affects the welfare of a rich man.

Stern assumes a value of almost zero for delta, that is, no time preference discount at all, on the grounds that the welfare of all future generations should be regarded as being as important here and now as that of the present generation (he raises it marginally to 0.1% to cover the possibility of the world coming to an end, in which case future generations would not exist), and a value of 1 for eta, implying that a loss (or gain) of, say, 10% of income for a poor man is precisely the same, in welfare terms, as a loss (or gain) of 10% for a rich man.

Most economists – and probably most laymen – would regard both these assumptions (in particular the first) as being unjustifiably low. Both assumptions are, of course, entirely subjective and prescriptive, and bear no relation to how we behave (or arguably, so far as delta is concerned, could rationally behave) in real life. The uncertainty about what overall discount rate Stern assumes relates to a lack of clarity about what he is assuming for the rate of future economic growth per head of population. But it would seem that he assumes an overall discount rate of 2.1% for the next hundred years, steadily declining thereafter, on the somewhat counterintuitive grounds that 'if *(sic)* uncertainty rises as we go into the future, this would work to reduce the discount rate' (N Stern, 2006, op. cit., p. 59).

3. P Dasgupta, 'Commentary: The Stern Review's economics of climate change', *National Institute Economic Review*, 199, 2007 (Sir Partha Dasgupta is Frank Ramsey Professor of Economics at Cambridge University); and W Nordhaus, 'The Stern Review on the economics of climate change', *Journal*

of Economic Literature, 45, no. 3, September 2007. See also W Nordhaus, *The Challenge of Global Warming: Economic models and environmental policy*, 2007. (William Nordhaus is Professor of Economics at Yale University.)

4. M Weitzman, 'The Stern Review of the economics of climate change', *Journal of Economic Literature*, 45, no. 3, September 2007.

5. See, for example, Nordhaus, 2007, op. cit..

6. D Lal, 'The Changing Climate II: Science, ethics, economics-II', *Business Standard*, 17 July 2007.

7. This appears to be the view now taken by Professor Lovelock.

8. N Stern, 2006, op. cit., p. 54.

9. D Hume, *A Treatise of Human Nature*, 1978, (Book 3, Part 2, Section 1), pp. 483-4.

10. I Little, article on 'Stern Philosophy', submitted to *The Financial Times*, 2007, but not published.

11. Charles Dickens, *Bleak House*, 2003. See esp. Chapter IV.

12. M Weitzman, 2007, op. cit..

13. See, for example, Martin Wolf's various articles on the subject in *The Financial Times*; also P Dasgupta, 'A challenge to Kyoto', *Nature*, 449, September 2007, pp. 143-4.

14. The House of Commons Select Committee on Science and Technology very sensibly recommended, in 2006, 'that the term "precautionary principle" should not be used', and that it should 'cease to be included in policy guidance'.

15. It is important to recognize that the insurance analogy, although frequently used, is an extremely inexact one. Insurance is a means of buying, in advance, compensation, in the event of an actuarially measurable risk occurring. The risk allegedly involved in global warming is neither actuarially measurable nor compensatable. Adaptation (which is how, as individuals, we deal with economic vicissitudes), 'mitigation' (that is, attempting to reduce the risk, which is how, collectively, we approach national security issues), and insurance, are three completely different ways of dealing with uncertainty. The nature of the uncertainty, and all the other facts of the case, will determine which approach is practicable and sensible for any given risk.

16. See, in this context, W Nordhaus, 2007, op. cit., Professor Nordhaus also observes that 'our empirical understanding of the economy-climate nexus is insufficiently rich to allow us to make precise judgments about the distribution or impact of extremely improbable events'.

17. M Rees, *Our Final Century: Will the human race survive the twenty-first century?*, 2004.

Chapter 8: Summary and Conclusion: A Convenient Religion

1. The media, of course, love scare stories of all kinds, as a means of

attracting the reader's or viewer's attention. The endless litany of medical scare stories is a case in point. Global warming, with its alleged existential threat to the planet, is the greatest scare story of all, and the media inevitably make the most of it. And just as the media welcome scare stories as a means of attracting the reader's or viewer's attention, so many scientists and others who should know better are tempted to speak in alarmist terms as a means of attracting the media's attention.

2. See, for example, R Lindzen, 'Climate of Fear', *Wall Street Journal*, 12 April 2006.

3. It is frequently overlooked, incidentally, that while technological progress may well, over time, reduce the cost of some forms of non-carbon energy, it may also do the same for carbon-based energy, in particular that based on coal.

4. P J Crutzen, 'Albedo enhancement by stratospheric sulfur injections: A contribution to resolve a policy dilemma?', *Climatic Change*, 77, nos. 3-4, August 2006, pp. 211-20.

5. According to Caldeira, 'If we could pour a five-gallon bucket's worth of sulfate particles per second into the stratosphere, it might be enough to keep the earth from warming for 50 years. Tossing twice as much up there could protect us into the next century', ('How to Cool the Globe', *New York Times,* 24 October 2007.) The cost of doing this, incidentally, is several orders of magnitude less than the cost of global decarbonization.

6. Unlike the Nobel Prize for Peace, which is simply an expression of political opinion, the Nobel Prizes for science are an objective recognition of achievement. The two awarding bodies are entirely different, too.

7. Needless to say this is strongly opposed by both Friends of the Earth and Greenpeace: further evidence of the politico-religious nature of their agenda. Professor Lovelock, by contrast, who has no such agenda but is seriously concerned at the prospect of extreme global warming, supports it.

8. Adam Smith, 1755. Published in A Smith, *Essays on Philosophical Subjects*, 1980.

9. M Wolf, 'Why emissions curb must be simple', *Financial Times*, 16 March 2007.

10. Rowan Williams, Archbishop of Canterbury. Transcript of Archbishop's interview on climate change with *The Today Programme*, 29 March 2006.

11. In fairness it should be noted that the Catholic Church's stance on this is more careful. In his 2008 New Year's Day message, Pope Benedict XVI declared that 'Human beings, obviously, are of supreme worth vis-à-vis creation as a whole. Respecting the environment does not mean considering material or animal nature more important than man ... Humanity today is rightly concerned about the ecological balance of tomorrow. It is important

for assessments in this regard to be carried out prudently, in dialogue with experts and uninhibited by ideological pressure to draw hasty conclusions'.

12. J A Schumpeter, *Capitalism, Socialism and Democracy*, 1952.

13. David Adam, 'Royal Society tells Exxon: stop funding climate change denial', *The Guardian,* 20 September 2006.

14. 'Some economists have analyzed the potential to use the international trade regime to respond to significant differences in the level of carbon prices applied in different economies. Countries could in theory impose a border tax on imports from countries with lower carbon prices ... There is a clear logic here.' (N Stern, 2006, op. cit., p. 551. While it decides that, on balance, this might not be a very good idea, the review feebly concludes that, 'Nevertheless, there remains the risk that in the face of significant and long-running divergences in the level of carbon prices across borders, industry will lobby for the introduction of these measures.' (p. 552).

15. 'Sarkozy warns China of carbon tariffs', http://.www.FT.com, 27 November 2007. Mr Sarkozy had earlier urged the President of the European Commission, in October 2007, to discuss, in the next six months, the implications of 'unfair competition' by firms outside the EU which do not have to abide by 'strict' European standards for CO_2 emissions.

Bibliography

Adam, D. 'Royal Society tells Exxon: stop funding climate change denial', *The Guardian,* 20 September 2006

Akasofu, Dr S-I. Written Testimony before the United States Senate Committee on Commerce, Science and Transportation Subcommittee on global climate change and impacts hearing on the projected and past effects of climate change: a focus on marine and terrestrial systems, 26 April 2006, Washington DC

Anderson, T.L. et al. 'Climate forcing by aerosols – a hazy picture' *Science* 300, 16 May 2003, pp.1103–4

Beckerman, W. & Hepburn, C. 'Ethics of the Discount Rate in the Stern Review on the Economics of Climate Change', *World Economics* 8 (1), January–March 2007

Black, R. 'Hurricanes and global warming - a link?', BBC News, 23 September 2005. Available at http://news.bbc.co.uk/1/hi/sci/tech/4276242.stm

Bond, G. et al. 'Persistent solar influence on North Atlantic climate during the Holocene', *Science* 294, 2001, pp. 2130–6

Booker, C. & North, R. *Scared to Death: The Anatomy of a Modern Madness*. London: Continuum, 2007

Botkin, D. 'Global Warming Delusions', *Wall Street Journal,* 17 October 2007

British Government. 'Iraq's Weapons of Mass Destruction', 24 September 2002

Brown A.G. et al., 'Roman vineyards in Britain: stratigraphic and palynological data from Wollaston in the Nene Valley, England', *Antiquity* 75 (290), 2001, pp. 745–57

Byatt, I. et al. 'The Stern Review: A Dual Critique – Economic Aspects', *World Economics* 7 (4), 2006, pp.199–229

Caldeira, K. 'How to Cool the Globe', *New York Times,* 24 October 2007

CAPE Last Interglacial Project Members, 'Last interglacial Arctic warmth confirms polar amplification of climate change', *Quaternary Science Reviews* 25, 2006, pp. 1383–1400

Carter, R.M. et al. 'The Stern Review: A dual critique – the science', *World Economics* 7 (4), 2006, pp.167–98

Chang, A. 'China answers emissions critics', *Associated Press*, 22 June 2007

Castles, I. & Henderson, D. 'The IPCC emissions scenarios: and economic- statistical critique', *Energy and Environment*, 14 (2&3), 2003, pp. 159–86

Chavez, F.P. et al. 'From anchovies to sardines and back: multidecadal change in the Pacific Ocean', *Science* 299, 2003, pp. 217–21

Copenhagen Consensus Center, Copenhagen Consensus Center [online: web] http://www.copenhagenconsensus.com

Crutzen, P.J. 'Albedo enhancement by stratospheric sulfur injections: A contribution to resolve a policy dilemma?', *Climatic Change* 77 (3&4), August 2006, pp. 211–20

Cunningham, S. et al. 'Temporal variability of the Atlantic Meridional Overturning Circulation at 26.5 ºN', *Science* 317, August 2007, pp. 935-8

Cutler, A. 'The Little Ice Age: When global cooling gripped the world', *Washtington Post,* 13 August 1997. Available at http://www-earth.usc.edu/geol150/evolution/images/littleiceage/ LittleIceAge.htm

Darby, D. et al. 'New Record Shows Pronounced Changes in Arctic Ocean Circulation and Climate', *EOS Transactions,* 82 (49), 2001, pp. 601–7

Darling, A. *House of Commons Daily Hansard*, 23 May 2007, Col. 1289. Available at http://www.publications.parliament.uk/pa/cm200607/cmhansrd/cm070523/debtext/70523-0005.htm#07052360001536

Dasgupta, P. *Comments on the Stern Review's Economics of Climate Change*. Cambridge: University of Cambridge, 2006

Dasgupta, P. 'A challenge to Kyoto', *Nature* 449, September 2007, pp. 143–4

Department for Business, Enterprise and Regulatory Reform, 'Draft options paper on renewables target', leaked to *The Guardian*, 13 August 2007

Department for Trade and Industry, Energy White Paper, 'Our Energy Future – Creating a Low Carbon Economy', 2003. Available at http://www.dti.gov.uk/files/file10719.pdf

Dickens, C. *Bleak House*. London: Penguin Books, Revised Edition, 2003

Ehrlich, P. *The Population Bomb*. New York: Ballantine, 1968

Environmental Protection Agency, *US Inventory of Greenhouse Gas Emissions and Sinks 1990–2005*, 2007. Available at http://www.epa.gov/climatechange/emissions/downloads/ 2007GHGFastFacts.pdf

European Environment Agency, *Annual European Community greenhouse gas inventory 1990–2005 and inventory report*. Submission to the UNFCCC Secretariat. EEA Technical Report No 7, 2007, available at http://reports.eea.europa.eu/technical_report_2007_7/en/Annual%20European%20Community%20greenhouse%20gas%20inventory%201990-2005%20and%20inventory%20report%202007.pdf

Gallo, K.P. et al. 'Temperature trends of the historical climatology network based on satellite-designated land use/land cover', *Journal of Climate* 12, 1999, pp. 1344–8

Gedney, N. et al. 'Detection of a direct carbon dioxide effect in continental river runoff records', *Nature* 439, 16 February 2006, pp. 835–8

Goddard Institute for Space Studies, NASA, 'Datasets and images – GISS surface temperature analysis.' Available at http://data.giss.nasa.gov/ gistemp/

Goklany, I.M. 'A Climate Policy for the Short and Medium Term: Stabilization or Adaptation?', *Energy & Environment* 16, 2005, pp. 667–80

Goklany, I.M. *The Improving State of Humanity*. Washington DC: Cato Institute, 2007

Goklany, I.M. 'Saving habitat and conserving biodiversity on a crowded planet', *BioScience* 48, 1998, pp. 941–953

Goklany, I.M. 'CO2 and biodiversity: Does the former affect the latter?', *CO2 Science* 5 (35), August 2002

Gore, A. *An Inconvenient Truth*. Emmaus, PA: Rodale, 2006

Gray, V. 'The truth about Tuvalu', *NZ Climate and Enviro Truth* 103, 15 June 2006

Greenhouse Gas Division of Environment Canada, *National Inventory Report, 1990–2005: Greenhouse Gas Sources and Sinks in Canada*, The Canadian Government's Submission to the UN Framework Convention on Climate Change, April 2007. Available at http://www.ec.gc.ca/pdb/ghg/inventory_report/2005_report/tdm-toc_eng.cfm

Gwynne, P. 'The Cooling World', *Newsweek*, 28 April 1975, (64). Available at http://brain-terminal.com/common/images.php/newsweek-cooling-world.pdf

Harvey, F., Bryant, C. & Aglionby, J. 'Producers, traders reap credits windfall', *Financial Times*, 26 April 2007. Available at http://www.ft.com/cms/s/c12930c0-f439-11db-88aa-000b5df10621.html

Hayek, F.A. 'The Economy, Science, and Politics', reprinted in *Studies in Philosophy, Politics and Economics*. London: Routledge & Kegan Paul, 1967

Helm, D. et al. 'Too good to be true? The UK's climate change record', 10 December 2007. Available at Http://www.dieterhelm.co.uk/publications/Carbon_record_2007.pdf

Henderson, P.D. 'Governments and climate change issues', *World Economics* 8 (2), April–June 2007

Hirschi, J. et al. 'Global warming and changes of continentality since 1948', *Weather* 82 (8), August 2007

Holgate, S.J. 'On the decadal rates of sea level change during the twentieth century', *Geophysical Research Letters* 34 (1), 4, January 2007

Holland, D. 'Bias and concealment in the IPCC process: the "Hockey-Stick" affair and its implications', *Energy and Environment* 18 (7&8), 2007

Houghton, J. *Evidence given to House of Lords Select Committee on Economic Affairs. Examination of Witnesses (Questions 40–59).* Available at http://www.publications.parliament.uk/pa/ld200506/ldselect/ldeconaf/12/5011803.htm

House of Commons Science and Technology Committee, 'Scientific advice, risk and evidence based policy making: Government response to the Committee's seventh report of session 2005–06', 27 February 2007, The House of Commons, The Stationery Office Limited

House of Lords Select Committee on Economic Affairs, 'The Economics of Climate Change.' 2nd Report of Session 2005-06. HL Paper 12-I available at http://www.publications.parliament.uk/pa/ld200506/ldselect/ldeconaf/12/12i.pdf. HL Paper 12-II (including oral and written evidence) available at http://www.publications.parliament.uk/pa/ld200506/ldselect/ldeconaf/12/1202.htm

Howard, L. *The Climate of London, Deduced From Meteorological Observations*. London: Phillips, 1818

Howell, D. & Nakhle, C. *Out of the Energy Labyrinth*. New York: I B Tauris, 2007

Hoyt, D.V. & Schatten, K.H. 'A discussion of plausible solar irradiance variations, 1700–1992', *Journal of Geophysical Research* 98, 1993, pp. 18895–906

Huang, S., Pollack, H.N. & Snen, P.O. 'Late Quaternary temperature changes seen in world-wide continental heat flow measurements', *Geophysical Research Letters* 24 (15), 1997, pp. 1947–50

Hulme, M. 'Chaotic world of climate truth', BBC News, 4 November 2006. Available at http://news.bbc.co.uk/1/hi/sci/tech/6115644.stm

Hume, D. *A Treatise of Human Nature*. Oxford: Oxford University Press, 1978 [1739]

International Energy Agency, *World Energy Outlook* 2007, OECD/IEA Paris, December 2007

IPCC, *Fourth Assessment Report: Climate Change 2007:* The Physical Science Basis. Summary for Policymakers. Cambridge: Cambridge University Press, 2007. Also available at http://www.ipcc.ch/ SPM2feb07.pdf

Jevrejeva, S., et al. 'Nonlinear trends and multiyear cycles in sea level records', *Journal of Geophysical Research* 111, 2006

Johannessen, O.M. et al. 'Recent Ice-Sheet Growth in the Interior of Greenland', *Sciencexpress*, 20 October 2005

Kai, M. 'China is shouldering its climate change burden', *Financial Times,* 4 June 2007

Kalnay E. & Cai, M. 'Impact of urbanization and land-use change on climate', *Nature* 423, 2003, pp. 528-. 531

Keatinge, W.R. 'Seasonal mortality among elderly people with unrestricted home heating', *British Medical Journal* 293, 1986, pp. 732–733

Keatinge, W.R. & Donaldson, G.C. 'The impact of global warming on health and mortality', *Southern Medical Journal* 97 (11), November 2004

Keatinge, W.R. et al. 'Heat related mortality in warm and cold regions of Europe: observational study', *British Medical Journal* 321, 2000, pp. 670–3

Keay, M. *The Dynamics of Power: Power Generation Investment in Liberalised Energy Markets*. Oxford: Oxford Institute for Energy Studies, 2006

Keigwin, L. 'The Little Ice Age and Mediaeval Warm Period in the Sargasso Sea', *Science* 274, 1996, pp. 1504–8

Khim, B.K. et al. 'Unstable climate oscillations during the late Holocene in the Eastern Bransfield Basin, Antarctic Peninsula', *Quaternary Research* 58, (234–245), 2002

Krabill, W. et al. 'Greenland ice sheet: high-elevation balance and peripheral thinning', *Science* 289, 2005, pp. 428–30

Laaidi, M. et al. 'Temperature-related mortality in France; a comparison between regions with different climates from the perspective of global warming', *International Journal of Biometeorology* 51 (2), November 2006, pp.145–53

Lal, D. 'Climate change: Ethics, science, economics – II', *Business Standard,* 17 July 2007

Lamb, H. *Climate, History and the Modern World*. London: Routledge, 1995 (2nd edition)

Lamb, H., 'The early Mediaeval warm epoch and its sequel', *Paleogeography, Paleoclimatology, Paleoecology* 1, 1965, pp.13–37

Landsea, C. FAQs: Consensus Statements by International Workshop on Tropical Cyclones-VI (IWTC-VI) Participants, Atlantic Oceanographic and Meteorological Laboratory, Hurricane Research Divison. Available at http://www.aoml.noaa.gov/hrd/tcfaq/G3.html

Laughton, M. 'Observations on the UKERC Report on *The Costs and Impacts of Intermittency*', featured as Appendix 2 of *Response of the Renewable Energy Foundation to the 2006 Energy Review Our Energy Challenge*, 13 April 2006. Available at http://www.dti.gov.uk/files/file30869.pdf

Lawson, N. *The View from No. 11: memoirs of a Tory radical.* London: Bantam Press, 1992

Lean, J., Beer, J. & Bradley, R.S. 'Reconstruction of solar irradiance since 1610: implications for climate change', *Geophysical Research Letters* 22, 1995, pp. 3195–8

Lindzen, R.S. 'Climate of Fear', *Wall Street Journal*, 12 April 2006

Lindzen, R.S., Chou, M-D., and Hou, A.Y., 'Does the earth have an adaptive infrared iris?', *Bulletin of the American Meteorological Society* 82 (3), 2001, pp. 417–32

Lockwood, M. 'A Rough Guide to Carbon Trading', *Prospect,* February 2007

Lomborg, B. *Cool It: The Skeptical Environmentalist's Guide to Global Warming*. New York: Alfred A Knopf, 2007

Lomborg, B. 'Perspective on Climate Change', Testimony to the US Senate's joint hearing of the Committee on Energy and Environment and the Committee on Science and Technology, 21 March 2007. Available at http://democrats.science.house.gov/Media/File/Commdocs/hearings/2007/energy/21mar/lomborg_testimony.pdf

Lomborg, B. (ed.) *Global Crises, Global Solutions: priorities for a world of scarcity*. Cambridge: Cambridge University Press, 2004

Lovelock, J. 'Ground Truths', *Prospect,* December 2007, p. 68

MacDonald, G.M, et al. 'Holocene treeline history and climate change across northern Eurasia', *Quaternary Research* 53, 2000, pp. 302–11

Malthus, R. *An Essay on the Principle of Population*. Oxford: Oxford World Classics, 1999 [1798]

Mann, M. et al., 'Global-scale temperature patterns and climate forcing over the past six centuries', *Nature* 392, 1998, pp. 779–87

Mann, M. et al., 'Northern hemisphere temperatures during the past millennium: inferences, uncertainties, and limitations', *Geophysical Research Letters* 26, 1999, pp. 759–62

McIntyre, S. Climate Audit [online: web] http://www.climateaudit.org/

McIntyre, S. & McKitrick, R. 'Corrections to the Mann et al. (1998) Proxy Data Base and Northern Hemisphere Average Temperature Series', *Energy and Environment* 14 (6), 1 November 2003, pp. 751–71

McIntyre, S. & McKitrick, R. 'Hockey sticks, principal components, and spurious significance', *Geophysical Research Letters* 32, 2005

McKitrick, R. et al. *Independent Summary for Policymakers – IPCC Fourth Assessment Report*. Vancouver, BC: The Fraser Institute, 2007

Meadows, D. et al. *The Limits to Growth.* New York: Universe Books, 1972

Met Office, News Release, 'Tropical Storms and Climate Change', 20 February 2006. Available at http://www.metoffice.com/corporate/pressoffice/2006/pr20060220.html

Met Office/Hadley Centre, Graph: 'Global Average Near-Surface Temperatures 1850–2006.' Last updated 2February 2007. Available at http://www.metoffice.gov.uk/research/hadleycentre/CR_data/Monthly/Hadplot_globe.gif

Met Office/Hadley Centre, 'Climate change and the greenhouse effect', December 2005

Millar, C.I. et al. 'Late Holocene forest dynamics, volcanism, and climate change at Whitewing Mountain and San Joaquin Ridge, Mono County, Sierra Nevada, CA, USA', *Quaternary Research* 66, 2006, p. 273–87

Morner, N.A. 'Estimating future sea level changes from past records', *Global and Planetary Change* 40, (1&2), January 2004, pp. 49–54

Morner, N.A., Tooley, M. & Possnert, G. 'New perspectives for the future of the Maldives', *Global and Planetary Change* 40, (1&2), January 2004, pp. 177–82

Nakicenovic, N., et al. *Special Report on Emissions Scenarios: A Special Report of Working Group III of the Intergovernmental Panel on Climate Change*. Cambridge: Cambridge University Press, 2000

National Climatic Date Center, State of the Climate yearly reports. Available at http://www.ncdc.noaa.gov/oa/climate/research/monitoring.html

National Hurricane Center, The deadliest hurricanes in the United States 1900–1996. Available at http://www.nhc.noaa.gov/pastdead.html

Naurzbaev, M.M., Hughes, M.K. & Vaganov, E.A. 'Tree-ring growth curves as sources of climatic information', *Quaternary Research* 62, 2004, p. 126–33

'Open Kyoto to debate: Sixty scientists call on Harper to revisit the science of global warming' *National Post*, Canada, 6 April 2006

Nedic, D. et al. *Security Assessment of Future UK Electricity Scenarios*. Norwich: Tyndall Centre for Climate Change Research, 2005, pp. 47–48

Newton, A., Thunell, R. & Stott, L. 'Climate and hydrographic variability in the Indo-Pacific Warm Pool during the last millennium', *Geophysical Research Letters* 33, 2006

Nordhaus, W. 'The Stern Review on the economics of climate change', *Journal of Economic Literature* 45 (3), September 2007

Nordhaus, W. *The Challenge of Global Warming: Economic models and environmental policy*. New Haven, CT: Yale University, 2007

North, G.D. 'Surface Temperature Reconstructions for the Last 2,000 Years', Statement before the Subcommittee on Oversight and Investigations Committee on Energy and Commerce US House of Representatives, 19 July 2006. Available at http://dels.nas.edu/dels/rpt_briefs/Surface_Temps_final.pdf

OECD, Round Table on Sustainable Development, 'Biofuels: Is the cure worse than the disease?' September 2007. Available at http://www.oecd.org/dataoecd/15/46/39348696.pdf

Okonski, K. (ed.), *The Water Revolution: Practical Solutions to Water Scarcity*. London: International Policy Press, 2006

Open Europe, 'Europe's dirty secret: Why the EU Emissions Trading Scheme isn't working', 2007. Available at http://www.openeurope.org.uk

Paddock, W. & Paddock, P. *Famine – 1975!* Boston: Little, Brown & Co., 1967

Pielke, R., Jr. 'Mistreatment of the economic impacts of extreme events in the Stern Review', *Global Environmental Change*, 2007

Pielke, R. & Landsea, C. 'Damage trends in Atlantic hurricanes', *Natural Hazards Review*, 2007

Pimentel, D. 'Ethanol Fuels: Energy balance, economics, and environmental impacts are negative', *Natural Resources Research* 12 (2), June 2003. Available at http://www.energyjustice.net/ethanol/pimentel2003.pdf

Popper, K. *The Logic of Scientific Discovery*. London: Routledge & Kegan Paul, 1959; first published in German in 1934 as *Die Logik der Forschung*

Raina, V. & Sangewar, C. 'The Siachen glacier', *Journal of the Geological Society of India* 70 (1), July 2007, pp.11–6

Ramanathan V. et al. 'Warming trends in Asia amplified by brown cloud solar absorption', *Nature* 448, 2007, pp. 575–8

Ramsey, F. 'A mathematical theory of saving', *Economic Journal* 38, 1928, pp. 543–59

Rees, M. *Our Final Century: Will the human race survive the twenty-first century?* London: Arrow Books, 2004, (2nd Edition)

Reiter, P. 'Malaria in England in the Little Ice Age', *Emerging Infectious Diseases*, 6 (1), Jan–Feb 2000

Richey, J.N. et al. '1400 year multiproxy record of climate variability from the northern Gulf of Mexico', *Geology* 35, 2007, pp. 423–426

Righelato, R. & Spracklen, D. 'Carbon mitigation by biofuels or by saving and restoring forests?' *Science* 317, August 2007, p. 902

Roberts, D., Manguin, S. & Mouchet, J. 'DDT house spraying and re-emerging malaria', *Lancet* 356, 2000, pp. 330–2

Schott, F. et al. *Geophysical Research Letters* 33, 2006

Schumpeter, J.A. *Capitalism, Socialism and Democracy*. London: Allen & Unwin, 1952, (4th edition)

Seager, S. et al. 'Is the Gulf Stream responsible for Europe's mild winters?' *Quarterly Journal of the Royal Meteorological Society* 128 (586), October 2002

Sense about Science, *Making Sense of the Weather and Climate*. London: Sense About Science, March 2007

Simpson, D. 'Tilting at windmills: The economics of wind power', David Hume Institute Occasional Paper No. 65, April 2004. Available at http://www.davidhumeinstitute.com/DHI%20Website/publications/hop/Wind%20Power%20paper.pdf

Simpson, D. et al. 'Influence of biomass burning during recent fluctuations in the slow growth of global tropospheric methane', *Geophysical Research Letters* 33 (22), 2006

Smith, A. *An Inquiry Into the Nature and Cause of the Wealth of Nations*. Oxford: Clarendon Press, 1976

Smith, A. *Essays on Philosophical Subjects*. Oxford: Clarendon, 1980

Smith, D. et al. 'Improved Surface Temperature Prediction for the Coming Decade from a Global Climate Model', *Science* 317, August 2007, pp.796–9

Solanki, S.K. & Fligge, M. 'Solar irradiance since 1874 revisited', *Geophysical Research Letters* 25, 1998, pp. 341–4

Soon, W. & Baliunas, S. 'Proxy climatic and environmental changes of the past 1000 years', *Climate Research* 23, 2003, pp. 89–110

Stern, N. *The Economics of Climate Change: The Stern Review*. Cambridge: Cambridge University Press, 2007

Stern, N. 'Reply to Byatt et al.' *World Economics* 7 (2) Apr–Jun 2006

Stern, N. 'What is the Economics of Climate Change?' *World Economics* 7 (2), Apr–Jun 2006

Svensmark, H. & Calder, N. *The Chilling Stars: A New Theory of Climate Change*. Cambridge: Icon Books, 2007

Svensmark, H., Pedersen, J. et al. 'Experimental evidence for the role of ions in particle nucleation under atmospheric conditions', *Proceedings of the Royal Society* A, London, October 2006

Thatcher, M. *The Downing Street Years*. London: Harper Collins, 1993

Tren, R. & Bate, R. *Malaria and the DDT Story*. London: Institute of Economic Affairs, 2001

Trenberth, K. 'Global Warming and Forecasts of Climate Change', *Climate Feedback: The climate change blog*, July 2007 [online:web] http://blogs.nature.com/climatefeedback/recent_contributors/kevin_trenberth/

Trenberth, K. 'Predictions of Climate', *Climate Feedback: The climate change blog*, June 2007 [online:web] http://blogs.nature.com/climatefeedback/ recent_contributors/ kevin_trenberth/

United Nations, *Kyoto Protocol to the United Nations Framework Convention on Climate Change*, 1997

United Nations Environment Programme, *Montreal Protocol on Substances that Deplete the Ozone Layer*, 1987

United Nations, Department of Economic and Social Affairs, Population Division [online: web] http://www.un.org/esa/population/unpop.htm

Van de Veer, J. 'High hopes and hard truths dictate future', *The Times*, 25 June 2007

Vincent, C. et al. 'Very high-elevation Mont Blanc glaciated areas not affected by the 20th century climate change', *Journal of Physical Research* 112, 2007

Vinther, B.M. et al. 'Extending Greenland temperature records into the late eighteenth century', *Journal of Geophysical Research* 3, 2006

Watson, R.T. et al. *IPCC. Climate change 2001: Synthesis Report. A contribution of Working Groups I, II and III to the Third Assessment Report of the Intergovernmental Panel on Climate Change. Cambridge: Cambridge University Press 2001, p. 398*

Wegman, E.J., Scott D.W. & Yasmin H. 'Ad Hoc Committee Report on the "Hockey Stick" Global Climate Reconstruction', 14 July 2006. Available at http://energycommerce.house.gov/108/home/07142006_Wegman_Report.pdf

Weitzman, M. 'The Stern Review of the economics of climate change', *Journal of Economic Literature* 45 (3), September 2007, pp.703–24

Whitehouse, D. 'Will sun's low activity arrest global warming?', *The Independent,* 5 December 2007

Whitehouse, D. *The Sun: A Biography*. Hoboken, NJ: John Wiley & Sons, 2004

Wingham, D. et al. 'Mass balance of the Antarctic ice sheet', *Philosophical Transactions of the Royal Society* A 364, 2006, pp.1627–35

Wunsch, C. 'Gulf Stream safe if wind blows and earth turns', *Nature* 428, 2004

Wunsch, C. Letters page, *The Economist*, 30 September 2006. Available at http://www.economist.com/opinion/displaystory.cfm?story_id=7963571

Zhen-Shan, L. & Xian, S. 'Multi-scale analysis of global temperature changes and trend of a drop in temperature in the next 20 years', *Meteorology and Atmospheric Physics* 95, 2007, pp. 115–21

Zwally, H.J. et al. 'Mass changes of the Greenland & Antarctic ice sheets & shelves & contributions to sea-level rise: 1992-2002', *Journal of Glaciology* 51 (175), December 2005, pp. 509–27

Index